Lecture Notes in Mathematics

Volume 2245

This series reports on new developments in all areas of mathematics and their applications - quickly, informally and at a high level. Mathematical texts analysing new developments in modelling and numerical simulation are welcome. The type of material considered for publication includes:

1. Research monographs
2. Lectures on a new field or presentations of a new angle in a classical field
3. Summer schools and intensive courses on topics of current research.

Texts which are out of print but still in demand may also be considered if they fall within these categories. The timeliness of a manuscript is sometimes more important than its form, which may be preliminary or tentative.

More information about this series at http://www.springer.com/series/304

Minh Kha • Peter Kuchment

Liouville-Riemann-Roch Theorems on Abelian Coverings

 Springer

Minh Kha
Department of Mathematics
University of Arizona
Tucson, AZ, USA

Peter Kuchment (iD)
Department of Mathematics
Texas A&M University
College Station, TX, USA

ISSN 0075-8434 ISSN 1617-9692 (electronic)
Lecture Notes in Mathematics
ISBN 978-3-030-67427-4 ISBN 978-3-030-67428-1 (eBook)
https://doi.org/10.1007/978-3-030-67428-1

Mathematics Subject Classification: Primary: 35A53, 35P99, 58J05; Secondary: 35J99, 35Q40, 19L10

This Springer imprint is published by the registered company Springer Nature Switzerland AG.
The registered company address is: Gewerbestrasse 11, 6330 Cham, Switzerland

*Dedicated to the memory of dear friends
and wonderful mathematicians
Misha Boshernitzan and Misha Shubin*

Preface

Counting the number (i.e., dimension of the space) or even confirming the existence of solutions of an elliptic equation on a compact manifold (or in a bounded domain in $\mathbb{R}^n$) is usually a rather impossible task, unstable with respect to small variations of parameters. On the other hand, the Fredholm index of the corresponding operator, as it was conjectured by I. M. Gel'fand [25] and proven by M. F. Atiyah and I. M. Singer [6, 7, 8, 9, 10], can be computed in topological terms. In particular, if the index of an operator L happens to be positive, this implies non-triviality of its kernel.

One might also be interested in index formulas in the case when the solutions are allowed to have some prescribed poles and have to have some mandatory zeros. Probably, the first result of this kind was the centuries old classical Riemann-Roch theorem [62, 63], which in an appropriate formulation provides the index of the $\bar{\partial}$-operator on a compact Riemannian surface, when a divisor of zeros and poles is provided. Analogs and extensions of this result were provided by V. G. Maz'ya and B. A. Plamenevskii [53] for elliptic boundary problems in domains and by N. S. Nadirashvili [57] for the Laplace-Beltrami operator on a complete compact.[1] Riemannian manifold with a prescribed divisor. The latter result has been generalized by M. Gromov and M. A. Shubin [29, 30, 31] to computing indices of elliptic operators in vector bundles over compact (as well as non-compact) manifolds, when a divisor mandates a finite number of zeros and allows a finite number of poles of solutions.

On the other hand, Liouville type theorems count the number of solutions that allow to have a "pole at infinity." Solution of an S.-T. Yau's problem [74, 75], given by T. H. Colding and W. P. Minicozzi II [16, 17, 18, 48], shows that on a Riemannian manifold of nonnegative Ricci curvature, the spaces of harmonic functions of fixed polynomial growths are finite dimensional. The result also applies to the Laplace-Beltrami operator on a nilpotent covering of a compact Riemannian manifold. No explicit formulas for these dimensions are available. However, an interesting

---

[1] A version for non-compact manifolds with "appropriate" conditions at infinity was also given.

case has been discovered by M. Avellaneda and F.-H. Lin [11] and J. Moser and M. Struwe [56]. It pertains periodic elliptic operators of divergent type, where exact dimensions can be computed and coincide with those for the Laplacian. This study has been extended by P. Li and J. Wang [49, 50] and brought to its natural limit in the case of periodic elliptic operators on co-compact abelian coverings by P. Kuchment and Y. Pinchover [43, 44].

A comparison of the works of Nadirashvili and Gromov and Shubin (N-G-S) and of Kuchment and Pinchover shows significant similarities in the techniques, as well as appearance of the same combinatorial expressions in the answers. Thus, a natural idea was considered that possibly the results could be combined somehow in the case of co-compact abelian coverings, if the infinity is "added to the divisor." In fact, the results of Nadirashvili, Gromov, and Shubin allowed having some conditions at infinity, if these lead to some kind of "Fredholmity." When one considers polynomial growth conditions at infinity, i.e., Liouville property, the results of [43, 44] show that such "Fredholmity" can be expected at the edges of the spectrum (in more technical terms, when the Fermi surface is discrete). Outside of the spectrum, this also works, vacuously. On the other hand, inside the spectrum, the space of polynomially growing solutions is infinitely dimensional, which does not leave much hope for Liouville-Riemann-Roch type results. We thus concentrate on the spectral edge case (the results of [11, 56] correspond to the case of the bottom of the spectrum).

This work shows that such results indeed can be obtained, although they come out more intricate than a simple-minded expectation would suggest. Namely, the interaction between the finite divisor and the point at infinity turns out to be non-trivial.

The structure of the text is as follows: Chap. 1 contains a survey of the required information about periodic elliptic operators, Liouville type theorems (including some new observations), and Nadirashvili-Gromov-Shubin version of the Riemann-Roch theorem. Chapter 2 contains formulations of the main results of Liouville-Riemann-Roch type. In some cases, the Riemann-Roch type equalities cannot be achieved (counterexamples exist), while inequalities still hold. These inequalities, however, can be applied, the same way the equalities are, for proving the existence of solutions of elliptic equations with prescribed zeros, poles, and growth at infinity. Proofs of these results are mostly contained in Chap. 3, while proofs of some technical auxiliary statements are delegated to Chap. 5. Chapter 4 provides applications to some specific examples. The work ends with some final remarks and conclusions.

College Station, TX, USA   Minh Kha
12/2019   Peter Kuchment

Acknowledgements

The work of the authors was partially supported by NSF DMS grants. The authors express their gratitude to NSF for the support. M. Kha also thanks AMS and the Simons foundation for their travel grant support. Thanks also go to V. G. Maz'ya, Y. Pinchover and M. A. Shubin for the insightful discussions on the topic, and the reviewers for carefully reading the manuscript and useful suggestions. The authors are thankful to the editorial board members and staff of *Lecture Notes in Mathematics* for their diligent work.

Contents

1   **Preliminaries** 1
   1.1   Periodic Elliptic Operators on Abelian Coverings..................... 1
   1.2   Floquet Transform 4
   1.3   Bloch and Fermi Varieties 5
   1.4   Floquet-Bloch Functions and Solutions............................ 6
   1.5   Liouville Theorem on Abelian Coverings 7
   1.6   Some Properties of Spaces $V_N^p(A)$ 9
   1.7   Explicit Formulas for Dimensions of Spaces $V_N^\infty(A)$.................. 10
   1.8   The Nadirashvili-Gromov-Shubin Version of the
       Riemann-Roch Theorem for Elliptic Operators on
       Noncompact Manifolds 12
       1.8.1   Some Notions and Preliminaries 12
       1.8.2   Point Divisors 14
       1.8.3   Rigged Divisors... 15
       1.8.4   Nadirashvili-Gromov-Shubin Theorem
            on Noncompact Manifolds 17

2   **The Main Results** 23
   2.1   Non-empty Fermi Surface 23
       2.1.1   Assumptions 24
       2.1.2   Spaces 25
       2.1.3   Results... 26
   2.2   Empty Fermi Surface 30

3   **Proofs of the Main Results** 35
   3.1   Some Notions 35
   3.2   Proof of Theorem 2.2 35
   3.3   Proof of Theorem 2.5 41
       3.3.1   Proof of Proposition 2.7 42
   3.4   Proof of Theorem 2.8 43
       3.4.1   Proof of Proposition 2.9 44
       3.4.2   Proof of Proposition 2.10................................... 44

          3.4.3   Proof of Proposition 2.11 47
          3.4.4   Proof of Corollary 2.12 49
    3.5   Proof of Theorem 2.16 49

4   **Specific Examples of Liouville-Riemann-Roch Theorems** 55
    4.1   Self-Adjoint Operators 55
          4.1.1   Periodic Operators with Non-degenerate Spectral Edges 57
          4.1.2   Periodic Operators with Dirac Points 60
    4.2   Non-Self-Adjoint Second Order Elliptic Operators 62

5   **Auxiliary Statements and Proofs of Technical Lemmas** 67
    5.1   Properties of Floquet Functions on Abelian Coverings 67
    5.2   Basic Properties of the Family $\{A(k)\}_{k \in \mathbb{C}^d}$ 69
    5.3   Properties of Floquet Transforms on Abelian Coverings 70
    5.4   A Schauder Type Estimate 73
    5.5   A Variant of Dedekind's Lemma 74
    5.6   Proofs of Some Other Technical Statements 75
          5.6.1   Proof of Theorem 1.13 75
          5.6.2   Proof of Theorem 1.14 82
          5.6.3   Proof of Corollary 1.27 84

A   **Final Remarks and Conclusions** 85

References 87

Index 91

Chapter 1
Preliminaries

Abstract This chapter contains a survey of the required information about abelian co-compact coverings and periodic elliptic operators on them. The function spaces that will be used throughout the book are introduced, as well as the main tools and notions of the so-called Floquet (or Floquet-Bloch) theory, in particular the notions of Floquet and Bloch functions. After that, spaces of polynomially growing (in an L_p-sense) solutions are discussed. Liouville type theorems (for $p = \infty$) due to works by Avellaneda and Lin, Moser and Struwe, and Kuchment and Pinchover, as well as their generalizations for $p \neq \infty$ are presented. Also, the results and some techniques of Nadirashvili and Gromov and Shubin on versions of Riemann-Roch theorem for elliptic operators are described.

1.1 Periodic Elliptic Operators on Abelian Coverings

Let X be a noncompact smooth Riemannian manifold of dimension n equipped with an isometric, properly discontinuous, free, and co-compact[1] action of a finitely generated Abelian discrete group G. We denote by $g \cdot x$ the action of an element $g \in G$ on $x \in X$.

Consider the (compact) orbit space $M := X/G$. We thus are dealing with a regular abelian covering π of a compact Riemannian manifold:

$$X \xrightarrow{\pi} M,$$

with G as its **deck group**.

Remark 1.1 Not much harm will be done, if the reader assumes that $X = \mathbb{R}^d$ and M is the torus $\mathbb{T}^d = \mathbb{R}^d/\mathbb{Z}^d$. The results are new in this case as well. The only warning is that in this situation the dimension of X and the rank of the group $\mathbb{Z}^d$ coincide, while this is not required in general, and d can be less or larger than n. As

---

[1] I.e., its quotient (orbit) space is compact.

© The Author(s), under exclusive license to Springer Nature Switzerland AG 2021
M. Kha, P. Kuchment, *Liouville-Riemann-Roch Theorems on Abelian Coverings*,
Lecture Notes in Mathematics 2245, https://doi.org/10.1007/978-3-030-67428-1_1

we will show, the distinction between the rank d of the deck group and dimension n of the covering manifold X pops up in some results.

Let μ_M be the Riemannian measure of M and μ_X be its lifting to X. Then μ_X is a G-invariant Riemannian measure on X. We denote by $L^2(X)$ the space of L^2-functions on X with respect to μ_X.

We also consider the G-invariant bilinear[2] duality

$$\langle \cdot, \cdot \rangle : C_c^\infty(X) \times C^\infty(X) \to \mathbb{C}, \quad \langle f, g \rangle = \int_X f(x)g(x)\mathrm{d}\mu_X. \tag{1.1}$$

It extends by continuity to a G-invariant bilinear non-degenerate duality

$$\langle \cdot, \cdot \rangle : L^2(X) \times L^2(X) \to \mathbb{C}. \tag{1.2}$$

Let A be an elliptic[3] differential operator of order m on X with smooth coefficients. We will be assuming that A is G-**periodic**, i.e., A commutes with the action of G on X. Then A can be pushed down to an elliptic operator A_M on M. Equivalently, A is the lifting of A_M to X. We will assume in most cases (except unfrequent non-self-adjoint considerations) that the operator A is bounded below.

The formal adjoint A^* (transpose with respect to the bilinear duality (1.1)) to A is also a periodic elliptic operator of order m on X.

Note that since G is a finitely generated abelian group, it is the direct sum of a finite abelian group and $\mathbb{Z}^d$, where d is the **rank** of the torsion free subgroup of G. One can always eliminate the torsion part of G by switching to a sub-covering $X \to X/\mathbb{Z}^d$. In what follows, without any effect on the results, we could replace M by the compact Riemannian manifold $X/\mathbb{Z}^d$ and thus, we can work with $\mathbb{Z}^d$ as our new deck group. Therefore,

we assume henceforth that $G = \mathbb{Z}^d$, where $d \in \mathbb{N}$.

The **reciprocal lattice** G^* for the deck group $G = \mathbb{Z}^d$ is $(2\pi\mathbb{Z})^d$ and we choose $B = [-\pi, \pi]^d$ as its fundamental domain (**Brillouin zone** in physics). The quotient $\mathbb{R}^d/G^*$ is a torus, denoted by $\mathbb{T}^{*d}$. So, G^*-periodic functions on $\mathbb{R}^d$ can be naturally identified with functions on the torus $\mathbb{T}^{*d}$.

For any **quasimomentum** $k \in \mathbb{C}^d$, let γ_k be the **character** of the deck group G defined as $\gamma_k(g) = e^{ik \cdot g}$ (a quasimomentum is defined modulo the reciprocal lattice). If k is real, γ_k is unitary and vice versa. Abusing the notations slightly, we will sometimes identify a unitary character γ_k, which belongs to the dual group $\mathbb{T}^d$ of $\mathbb{Z}^d$, with its quasimomentum $k \in B$.

---

[2]One can also consider the sesquilinear form to obtain the results.

[3]Here ellipticity means that the principal symbol of the operator A does not vanish on the cotangent bundle with the zero section removed $T^*X \setminus (X \times \{0\})$.

Definition 1.1 We denote by $L_k^2(X)$ the space of all γ_k-automorphic function $f(x)$ in $L_{loc}^2(X)$, i.e. such that

$$f(g \cdot x) = \gamma_k(g) f(x), \quad \text{for a.e } x \in X \text{ and } g \in G. \tag{1.3}$$

It is convenient at this moment to introduce, given a quasimomentum k, the following line (i.e., one-dimensional) vector bundle E_k over M:

Definition 1.2 Given any $k \in \mathbb{C}^d$, we consider the free left action of G on the Cartesian product $X \times \mathbb{C}$ (a trivial linear bundle over X) given by

$$g \cdot (x, z) = (g \cdot x, \gamma_k(g)z), \quad (g, x, z) \in G \times X \times \mathbb{C}.$$

Now E_k is defined as the orbit space of this action. Then the canonical projection $X \times \mathbb{C} \to M$ descends to the surjective mapping $E_k \to M$, thus defining a linear bundle E_k over M (see e.g., [47]).

Remark 1.2 The space $L_k^2(X)$ can be naturally identified with the space of L^2-sections of the bundle E_k.

This construction can be easily generalized to Sobolev spaces:

Definition 1.3 For a quasimomentum $k \in \mathbb{C}^d$ and a real number s, we denote by $H_k^s(X)$ the closed subspace of $H_{loc}^s(X)$ consisting of γ_k-automorphic (i.e., satisfying (1.3)) functions. Then $H_k^s(X)$ is a Hilbert space, when equipped with the natural inner product induced by the inner product in the Sobolev space $H^s(\overline{\mathcal{F}})$, where $\mathcal{F}$ is any fixed fundamental domain for the action of the group G on X.

Equivalently, the space $H_k^s(X)$ can be identified with the space $H^s(E_k)$ of all H^s-sections of the bundle E_k.

For any k, the periodic operator A maps continuously $H_k^m(X)$ into $L_k^2(X)$. This defines an elliptic operator $A(k)$ on the spaces of sections of the bundles E_k over the compact manifold M. When A is self-adjoint and k is real, the operator $A(k)$, with the space $H^m(E_k)$ as the domain, is an unbounded, bound below self-adjoint operator in $L^2(E_k)$. Thus its spectrum is discrete and eigenvalues can be labeled in non-decreasing order as

$$\lambda_1(k) \le \lambda_2(k) \le \cdots \to \infty. \tag{1.4}$$

A simple application of perturbation theory shows that $\lambda_j(k)$ are continuous piecewise-analytic G^*-periodic functions of k. Their ranges over the torus $\mathbb{T}^d$ are closed intervals I_j of the real axis, called **spectral bands**. The spectral bands tend to infinity, when $j \to \infty$ and may overlap (with any point being able to belong to at most finitely many bands), while they might leave some **spectral gaps** uncovered (e.g., [42, 73]).

1.2   Floquet Transform

Fourier transform is a major tool of studying linear constant coefficient PDEs, due to their invariance with respect to all shifts. The periodicity of the operator A suggests that it is natural to apply the Fourier transform with respect to the period group G to block-diagonalize A.

The group Fourier transform we have just mentioned is the so called **Floquet transform F** (see e.g., [20, 24, 41, 42]):

$$f(x) \mapsto \mathbf{F}f(k, x) = \sum_{g \in G} f(g \cdot x)\overline{\gamma_k(g)} = \sum_{g \in G} f(g \cdot x)e^{-ik \cdot g}, \quad k \in \mathbb{C}^d. \quad (1.5)$$

For reader's convenience, we collect some basic properties of Floquet transform in Chap. 5.

This transform, as one would expect, decomposes the original operator A on the non-compact manifold X into a direct integral of operators $A(k)$ acting on sections of line bundles E_k over k in torus $\mathbb{T}^d$:

$$A = \int_{\mathbb{T}^{*d}}^{\oplus} A(k)\mathrm{d}k, \quad L^2(X) = \int_{\mathbb{T}^{*d}}^{\oplus} L^2(E_k)\mathrm{d}k. \quad (1.6)$$

Here the measure $\mathrm{d}k$ is the normalized Haar measure on the torus $\mathbb{T}^{*d}$, which can be also considered as the normalized Lebesgue measure on the Brillouin zone B.[4]

One of the important consequences of this decomposition via the Floquet transform is the following well-known result, which in particular justifies the names "spectral band" and "spectral gap":

Theorem 1.4 ([40, 41, 61]) *The union of the spectra of operators $A(k)$ over the torus $\mathbb{T}^d$ is the spectrum of the periodic operator A: In other words, we have*

$$\sigma(A) = \bigcup_{k \in \mathbb{T}^d} \sigma(A(k)). \quad (1.8)$$

---

[4]In fact, we are mixing up the quasimomenta and characters here, and more appropriate formulas would be

$$A = \int_{B}^{\oplus} A(k)\mathrm{d}k, \quad L^2(X) = \int_{B}^{\oplus} L^2(E_k)\mathrm{d}k. \quad (1.7)$$

We will keep periodically abusing notations this way.

In the self-adjoint case, this can be rewritten as

$$\sigma(A) = \bigcup_{j \in \mathbb{N}} I_j = \bigcup_{j \in \mathbb{N}, k \in \mathbb{T}^d} \{\lambda_j(k)\}, \tag{1.9}$$

where $\lambda_j(k)$ are the eigenvalues of the operator $A(k)$, listed in non-decreasing order, and the finite closed segment I_j is the range of the function $\lambda_j(k)$.

1.3 Bloch and Fermi Varieties

We now recall a notion that plays a crucial role in studying periodic PDEs (see e.g., [23, 41, 42]).

Definition 1.5

(a) The (complex) **Bloch variety** B_A of the operator A consists of all pairs $(k, \lambda) \in \mathbb{C}^{d+1}$ such that λ is an eigenvalue of the operator $A(k)$.

$$B_A = \{(k, \lambda) \in \mathbb{C}^{d+1} : \lambda \in \sigma(A(k))\}.$$

Thus, the Bloch variety can be seen as the graph of the multivalued function $\lambda(k)$, which is also called the **dispersion relation**.[5]

(b) The (complex) **Fermi surface** $F_{A,\lambda}$ of the operator A at the energy level $\lambda \in \mathbb{C}$ consists of all quasimomenta $k \in \mathbb{C}^d$ such that the equation $A(k)u = \lambda u$ has a nonzero solution.[6] Equivalently, the Fermi surface $F_{A,\lambda}$ is the λ-level set of the dispersion relation. By definition, $F_{A,\lambda}$ is G^*-periodic.

(c) We denote by $B_{A,\mathbb{R}}$ and $F_{A,\lambda,\mathbb{R}}$ the **real Bloch variety** $B_A \cap \mathbb{R}^{d+1}$ and the **real Fermi surface** $F_{A,\lambda} \cap \mathbb{R}^d$, respectively.

(d) Whenever $\lambda = 0$, we will write F_A and $F_{A,\mathbb{R}}$ instead of $F_{A,0}$ and $F_{A,0,\mathbb{R}}$, correspondingly. This is convenient, since being at the spectral level λ, we could consider the operator $A - \lambda I$ instead of A and thus, $F_{A,\lambda} = F_{A-\lambda}$ and $F_{A,\lambda,\mathbb{R}} = F_{A-\lambda,\mathbb{R}}$. In other words, we will be able to assume w.l.o.g. that $\lambda = 0$.

Some important properties of Bloch variety and Fermi surface are stated in the next proposition (see e.g., [41, 42, 44]).

---

[5]In other words, it is the graph of the function $k \mapsto \sigma(A(k))$.

[6]In physics, the name "Fermi surface" is reserved only to a specific value of energy λ_F, called **Fermi energy** [4]. For other values of λ, the names **equi-energy**, or **constant energy surface** are used. For our purpose, the significance of the Fermi energy evaporates, and so we extend the name Fermi surface to all (even complex) values of λ.

Proposition 1.6

(a) *The Fermi surface and the Bloch variety are the zero level sets of some entire*
    (G^*-periodic in k) functions of finite orders on $\mathbb{C}^d$ and $\mathbb{C}^{d+1}$ respectively.
(b) *The Bloch (Fermi) variety is a G^*-periodic, complex analytic, codimension one*
    subvariety of $\mathbb{C}^{d+1}$ (correspondingly $\mathbb{C}^d$).
(c) *The real Fermi surface $F_{A,\lambda}$ either has zero measure in $\mathbb{R}^d$ or coincides with*
    the whole $\mathbb{R}^d$.
(d) *$(k, \lambda) \in B_A$ if and only if $(-k, \bar{\lambda}) \in B_{A^*}$. In other words, $F_{A,\lambda} = -F_{A^*,\bar{\lambda}}$ and*
    $F_{A,\mathbb{R}} = -F_{A^*,\mathbb{R}}$.

The Fermi and Bloch varieties encode much of crucial spectral information about
the periodic elliptic operator. For example, the absolute continuity of the spectrum
of a self-adjoint periodic elliptic operator (which is true for a large class of periodic
Schrödinger operators) can be reformulated as the absence of flat components in
its Bloch variety, which is also equivalent to a seemingly stronger fact that the real
Fermi surface at each energy level has zero measure (due to Proposition 1.6).

1.4   Floquet-Bloch Functions and Solutions

In this section, we introduce the notions of Bloch and Floquet solutions of periodic
PDEs and then state the Liouville theorem of [44].

Definition 1.7 For any $g \in G$ and quasimomentum $k \in \mathbb{C}^d$, we denote by $\Delta_{g;k}$
the "k-twisted" version of the first difference operator acting on functions on the
covering X as follows:

$$\Delta_{g;k}u(x) = e^{-ik \cdot g}u(g \cdot x) - u(x). \tag{1.10}$$

The iterated "twisted" finite differences of order N with quasimomentum k are
defined as

$$\Delta_{g_1,\dots,g_N;k} = \Delta_{g_1;k} \dots \Delta_{g_N;k}, \quad \text{for} \quad g_1, \dots, g_N \in G. \tag{1.11}$$

Definition 1.8 A function u on X is a **Floquet function of order N with quasi-**
momentum k if any twisted finite difference of order $N + 1$ with quasimomentum k
annihilates u. Also, a **Bloch function with quasimomentum k** is a Floquet function
of order 0 with quasimomentum k.

According to this definition, a Bloch function $u(x) \in L^2_{loc}(X)$ with quasimomentum
k is a γ_k-automorphic function on X, i.e., $u(g \cdot x) = e^{ik \cdot g}u(x)$ for any $g \in G$. If
the quasimomentum is *real*, then for any compact subset K of X, the sequence
$\{\|u\|_{L^2(gK)}\}_{g \in G}$ is bounded (in fact, constant), i.e., belongs to $\ell^\infty(G)$.

It is also known [44] that $u(x)$ is a Floquet function of order N with quasimo-mentum k if and only if u can be represented in the form

$$u(x) = e_k(x) \left(\sum_{|j| \leq N} [x]^j p_j(x) \right),$$

where $j = (j_1, \ldots, j_d) \in \mathbb{Z}_+^d$, and the functions p_j are G-periodic. Here for any $j \in \mathbb{Z}^d$, we define

$$|j| := |j_1| + \ldots |j_d|, \tag{1.12}$$

while $e_k(x)$ and $[x]^j$ are analogs of the exponential e^{ikx} and the monomial x^j on $\mathbb{R}^d$ (see [44] for details). For convenience, in Sect. 5.3, we collect some basic facts of Floquet functions on abelian coverings. In the flat case $X = \mathbb{R}^d$, a Floquet function of order N with quasimomentum k is the product of the plane wave e^{ikx} and a polynomial of degree N with G-periodic coefficients.

An important consequence of this representation is that any Floquet function $u(x) \in L_{loc}^2(X)$ of order N with a real quasimomentum satisfies the following L^2-growth estimate

$$\|u\|_{L^2(gK)} \leq C(1 + |g|)^N, \quad \forall g \in G \quad \text{and} \quad K \Subset X.$$

Here $|g|$ is defined according to (1.12), and we have used the Švarc-Milnor lemma from geometric group theory (see e.g., [52, Lemma 2.8]) to conclude that on a Riemannian co-compact covering X, the Riemannian distance between any compact subset K and its g-translation gK is comparable with $|g|$.

If u is continuous, the above L^2-growth estimate can be replaced by the corresponding L^∞-growth estimate.

1.5 Liouville Theorem on Abelian Coverings

We now need to introduce the spaces of polynomially growing solutions of the equation $Au = \lambda u$.

To simplify the notations, we will assume from now on that $\lambda = 0$, since, as we discussed before, we can deal with the operator $A - \lambda$ instead of A (see also Definition 1.5 d).

Definition 1.9 Let $K \Subset X$ be a compact domain such that X is the union of all G-translations of K, i.e.,

$$X = \bigcup_{g \in G} gK. \tag{1.13}$$

For any s, $N \in \mathbb{R}$ and $1 \leq p \leq \infty$, we define the vector spaces

$$V_N^p(X) := \left\{ u \in C^\infty(X) \mid \{\|u\|_{L^2(gK)} \cdot \langle g \rangle^{-N}\}_{g \in G} \in \ell^p(G) \right\},$$

and

$$V_N^p(A) := \left\{ u \in V_N^p(X) \mid Au = 0 \right\}.$$

Here $\langle g \rangle := (1 + |g|^2)^{1/2}$.

It is not hard to show that these spaces are independent of the choice of the compact subset K satisfying (1.13). In particular, one can take as K a fundamental domain for G-action on X. Moreover, we have $V_{N_1}^{p_1}(X) \subseteq V_{N_2}^{p_2}(X)$ and $V_{N_1}^{p_1}(A) \subseteq V_{N_2}^{p_2}(A)$ whenever $N_1 \leq N_2$ and $p_1 \leq p_2$.

Definition 1.10 For $N \geq 0$, we say that the **Liouville theorem of order** (p, N) **holds for A**, if the space $V_N^p(A)$ is finite dimensional.

Now we can restate one of the main results in [44] as follows

Theorem 1.11 *[44]*

(i) *The following statements are equivalent:*

    (1) *The cardinality of the real Fermi surface $F_{A,\mathbb{R}}$ is finite modulo G^*-shifts, i.e., Bloch solutions exist for only finitely many unitary characters γ_k.*
    (2) *The Liouville theorem of order (∞, N) holds for A for some $N \geq 0$.*
    (3) *The Liouville theorem of order (∞, N) holds for A for all $N \geq 0$.*

(ii) *Suppose that the Liouville theorem holds for A. Then for any $N \in \mathbb{N}$, each solution $u \in V_N^\infty(A)$ can be represented as a finite sum of Floquet solutions:*

$$u(x) = \sum_{k \in F_{A,\mathbb{R}}} \sum_{0 \leq j \leq N} u_{k,j}(x), \tag{1.14}$$

where each $u_{k,j}$ is a Floquet solution of order j with a quasimomentum k.

(iii) *A crude estimate of the dimension of $V_N^\infty(A)$:*

$$\dim V_N^\infty(A) \leq \binom{d+N}{N} \cdot \sum_{k \in F_{A,\mathbb{R}}} \dim \operatorname{Ker} A(k) < \infty.$$

Due to the relation between $F_{A,\mathbb{R}}$ and $F_{A^*,\mathbb{R}}$ (see Proposition 1.6), the Liouville theorem holds for A if and only if it also holds for A^*.

1.6 Some Properties of Spaces $V_N^p(A)$

Notation 1.12 *For a real number r, we denote by $\lfloor r \rfloor$ the largest integer that is strictly less than r, while $[r]$ denotes the largest integer that is less or equal than r.*

The following statement follows from Theorem 1.11 (ii):

Lemma 1.1 $V_N^\infty(A) = V_{[N]}^\infty(A)$ *for any non-negative real number N.*

The proofs of the next two theorems are delegated to the Sect. 5.6.

Theorem 1.13 *For each $1 \le p < \infty$ such that $pN > d$, one has*

$$V_{\lfloor N-d/p \rfloor}^\infty(A) \subseteq V_N^p(A).$$

If, additionally, the Fermi surface $F_{A,\mathbb{R}}$ is finite modulo G^*-shifts, then

$$V_{\lfloor N-d/p \rfloor}^\infty(A) = V_N^p(A).$$

Corollary 1.1 *If the Fermi surface $F_{A,\mathbb{R}}$ is finite modulo G^*-shifts, then Liouville theorem of order (p, N) with $pN > d$ holds for A if and only if the Liouville theorem of order (∞, N) holds for A for some $N \ge 0$, and thus, according to Theorem 1.11, for all $N \ge 0$.*

The following theorem could be regarded as a version of the unique continuation property at infinity for the periodic elliptic operator A.

Theorem 1.14 *Assume that $F_{A,\mathbb{R}}$ is finite (modulo G^*-shifts). Then the space $V_N^p(A)$ is trivial if either one of the following conditions holds:*

(a) $p \ne \infty$, $pN \le d$.
(b) $p = \infty$, $N < 0$.

In fact, a more general version of these results holds:

Theorem 1.15 *Let $p \in [1, \infty)$. Let also ϕ be a continuous, positive function defined on $\mathbb{R}^+$ such that*

$$N_{p,\phi} := \sup\left\{ N \in \mathbb{Z} : \int_0^\infty \phi(r)^{-p} \cdot \langle r \rangle^{pN+d-1} \mathrm{d}r < \infty \right\} < \infty.$$

We define $\mathcal{V}_\phi^p(A)$ as the space of all solutions u of $A = 0$ satisfying the condition

$$\sum_{g \in \mathbb{Z}^d} \|u\|_{L^2(gK)}^p \cdot \phi(|g|)^{-p} < \infty$$

holds for some compact domain K satisfying (1.13).

If $F_{A,\mathbb{R}}$ is finite (modulo G^*-shifts), then one has

- *If $N_{p,\phi} \geq 0$, then $\mathcal{V}_\phi^p(A) = V_{N_{p,\phi}}^\infty(A)$.*
- *If $N_{p,\phi} < 0$, then $\mathcal{V}_\phi^p(A) = \{0\}$.*

Note that if $\phi(r) = \langle r \rangle^N$, then $\mathcal{V}_\phi^p(A) = V_N^p(A)$.

The proofs of Theorems 1.13 and 1.14 (provided in Sect. 5.6) easily transfer to this general version. We thus skip the proof of Theorem 1.15 (never used later on in this text), leaving this as an exercise for the reader.

Remark 1.3 It is worthwhile to note that results of this section did not require the assumption of discreteness of spectra of the operators $A(k)$. This is useful, in particular, when considering overdetermined problems.

1.7   Explicit Formulas for Dimensions of Spaces $V_N^\infty(A)$

In order to describe explicit formulas for the dimensions of $V_N^\infty(A)$, we need to introduce some notions from [44]. Recall that, for each quasimomentum k, $A(k)$ belongs to the space $\mathcal{L}(H_k^m(X), L_k^2(X))$ of bounded linear operators acting from $H_k^m(X)$ to $L_k^2(X)$. For a real number s, the spaces $H_k^s(X)$ are the fibers of the following analytic G^*-periodic Hilbert vector bundle over $\mathbb{C}^d$:

$$\mathcal{E}^s := \bigcup_{k \in \mathbb{C}^d} H_k^s(X) = \bigcup_{k \in \mathbb{C}^d} H^s(E_k). \tag{1.15}$$

Consider a quasimomentum k_0 in $F_{A,\mathbb{R}}$. We can locally trivialize[7] the vector bundle $\mathcal{E}^s$, so that in a neighborhood of k_0, $A(k)$ becomes an analytic family of bounded operators from $H_{k_0}^s$ to $L_{k_0}^2$ (see Sect. 5.2). Suppose that the spectra of operators $A(k)$ are discrete for any value of the quasimomentum k.

Assume now that zero is an eigenvalue of the operator $A(k_0)$ with algebraic multiplicity r. Let Υ be a contour in $\mathbb{C}$ separating zero from the rest of the spectrum of $A(k_0)$. According to the perturbation theory (see Proposition 5.3), we can pick a small neighborhood of k_0 such that the contour Υ does not intersect with $\sigma(A(k))$ for any k in this neighborhood. We denote by $\Pi(k)$ the r-dimensional Riesz spectral projector [33, Ch. III, Theorem 6.17] for the operator $A(k)$, associated with the contour Υ. Now one can pick an orthonormal basis $\{e_j\}_{1 \leq j \leq r}$ in the range of $\Pi(k_0)$ and define $e_j(k) := \Pi(k)e_j$. Then let us consider the $r \times r$ matrix $\lambda(k)$ of the operator $A(k)\Pi(k)$ in the basis $\{e_j(k)\}$, i.e.,

$$\lambda_{ij}(k) = \langle A(k)e_j(k), e_i(k) \rangle = \langle A(k)\Pi(k)e_j, e_i \rangle. \tag{1.16}$$

---

[7]In fact, $\mathcal{E}^s$ is globally analytically trivializable (see e.g., [41, 76]) although we do not need this fact here.

Remark 1.4

- An important special case is when $r = 1$ near k_0. Then $\lambda(k)$ is just the band function that vanishes at k_0.
- It will be sometimes useful to note that our considerations in this part will not change if we multiply $\lambda(k)$ by an invertible matrix function analytic in a neighborhood of k_0.

Now, using the Taylor expansion around k_0, we decompose $\lambda(k)$ into the series of homogeneous matrix valued polynomials:

$$\lambda(k) = \sum_{j \geq 0} \lambda_j (k - k_0), \tag{1.17}$$

where each λ_j is a $\mathbb{C}^{r \times r}$-valued homogeneous polynomial of degree j in d variables.

For each quasimomentum $k_0 \in F_{A,\mathbb{R}}$, let $\ell_0(k_0)$ be the order of **the first non-zero term** of the Taylor expansion (1.17) around k_0 of the matrix function $\lambda(k)$.

The next result of [44] provides explicit formulas for dimensions of the spaces $V_N^\infty(A)$.

In order to avoid misunderstanding the formulas below, we adopt the following agreement:

Definition 1.16 If in some formulas throughout this text one encounters a binomial coefficient $\binom{A}{B}$, where $A < B$ (in particular, when $A < 0$) we define its value to be equal to zero.

Theorem 1.17 *[44] Suppose that the real Fermi surface $F_{A,\mathbb{R}}$ is finite (modulo G^*-shifts) and the spectrum $\sigma(A(k))$ is discrete for any quasimomentum k. Then,*

(a) For each integer $0 \leq N < \min\limits_{k \in F_{A,\mathbb{R}}} \ell_0(k)$, we have

$$\dim V_N^\infty(A) = \sum_{k \in F_{A,\mathbb{R}}} m_k \left[\binom{d+N}{d} - \binom{d+N-\ell_0(k)}{d} \right], \tag{1.18}$$

where m_k is the algebraic multiplicity of the zero eigenvalue of the operator $A(k)$.

(b) If for every $k \in F_{A,\mathbb{R}}$, $\det \lambda_{\ell_0(k)}$ is not identically equal to zero, formula (1.18) holds for any $N \geq 0$.

It is worthwhile to note that the positivity of $\ell_0(k)$ is equivalent to the fact that both algebraic and geometric multiplicities of the zero eigenvalue of the operator $A(k)$ are the same (which holds, e.g., in the self-adjoint case). Also, the non-vanishing of the determinant of $\lambda_{\ell_0(k)}$ implies that $\ell_0(k) > 0$.

1.8 The Nadirashvili-Gromov-Shubin Version of the Riemann-Roch Theorem for Elliptic Operators on Noncompact Manifolds

It will be useful to follow closely the paper [31] by M. Gromov and M. Shubin, addressing its parts that are relevant for our considerations.

1.8.1 Some Notions and Preliminaries

Through this section, P will denote a linear elliptic differential expression with smooth coefficients on a non-compact Riemannian manifold $\mathcal{X}$ (later on, $\mathcal{X}$ will be the space of an abelian co-compact covering). We denote by P^* its transpose (also an elliptic differential operator), defined via the identity

$$\langle Pu, v \rangle = \langle u, P^*v \rangle, \quad \forall u, v \in C_c^\infty(\mathcal{X}),$$

where $\langle \cdot, \cdot \rangle$ is the bilinear duality (1.1).

We notice that both P and P^* can be applied as differential expressions to any smooth function on $\mathcal{X}$ and these operations keep the spaces $C^\infty(\mathcal{X})$ and $C_c^\infty(\mathcal{X})$ invariant.

We assume that P and P^* are defined as operators on some domains Dom P and Dom P^*, such that

$$C_c^\infty(\mathcal{X}) \subseteq \text{Dom } P \subseteq C^\infty(\mathcal{X}), \tag{1.19}$$

$$C_c^\infty(\mathcal{X}) \subseteq \text{Dom } P^* \subseteq C^\infty(\mathcal{X}). \tag{1.20}$$

Definition 1.18 We denote by Im P and Im P^* the ranges of P and P^* on their corresponding domains, i.e.

$$\text{Im } P = P(\text{Dom } P), \quad \text{Im } P^* = P^*(\text{Dom } P^*). \tag{1.21}$$

As usual, Ker P and Ker P^* denote the spaces of solutions of the equations $Pu = 0$, $P^*u = 0$ in Dom P and Dom P^* respectively.

We also need to define some auxiliary spaces.[8] Namely, assume that we can choose linear subspaces $\mathrm{Dom}'P$ and $\mathrm{Dom}'P^*$ of $C^\infty(\mathcal{X})$ so that[9]

($\mathcal{P}1$)

$$C_c^\infty(\mathcal{X}) \subseteq \mathrm{Dom}'P \subseteq C^\infty(\mathcal{X}), \tag{1.22}$$

$$C_c^\infty(\mathcal{X}) \subseteq \mathrm{Dom}'P^* \subseteq C^\infty(\mathcal{X}), \tag{1.23}$$

and

($\mathcal{P}2$)

$$\mathrm{Im}\, P^* \subseteq \mathrm{Dom}'P, \quad \mathrm{Im}\, P \subseteq \mathrm{Dom}'P^*.$$

($\mathcal{P}3$) The bilinear pairing $\int_{\mathcal{X}} f(x)g(x)\mathrm{d}\mu_{\mathcal{X}}$ (see (1.1)) makes sense for functions from the relevant spaces, to define the pairings

$$\langle \cdot, \cdot \rangle : \mathrm{Dom}'P^* \times \mathrm{Dom}\, P^* \mapsto \mathbb{C}, \quad \langle \cdot, \cdot \rangle : \mathrm{Dom}\, P \times \mathrm{Dom}'P \mapsto \mathbb{C},$$

so that

($\mathcal{P}4$) The duality ("integration by parts formula")

$$\langle Pu, v \rangle = \langle u, P^*v \rangle, \quad \forall u \in \mathrm{Dom}\, P, \; v \in \mathrm{Dom}\, P^*$$

holds.

Remark 1.5 The notation Dom' might confuse the reader, leading her to thinking that this is a different domain of the same differential expression. It is rather an object **dual** to the domain Dom.

We also need an appropriate notion of a polar (annihilator) to a subspace:

Definition 1.19 For a subspace $L \subset \mathrm{Dom}\, P$, its **annihilator** L° is the subspace of $\mathrm{Dom}'P$ consisting of all elements of $\mathrm{Dom}'P$ that are orthogonal to L with respect to the pairing $\langle \cdot, \cdot \rangle$:

$$L^\circ = \{u \in \mathrm{Dom}'P \mid \langle v, u \rangle = 0, \text{ for any } v \in L\}.$$

Analogously, M° is the annihilator in $\mathrm{Dom}'P^*$ of a linear subspace $M \subset \mathrm{Dom}\, P^*$ with respect to $\langle \cdot, \cdot \rangle$.

---

[8]Most of the complications in definitions here and below come from non-compactness of the manifold.

[9]A variety of examples is provided in further chapters.

Following [31], we now introduce an appropriate for our goals notion of Fredholm property.

Definition 1.20 The operator P, as above, is a **Fredholm operator on** $\mathcal{X}$ if the following requirements are satisfied:

(i)

$$\dim \operatorname{Ker} P < \infty, \dim \operatorname{Ker} P^* < \infty$$

and
(ii)

$$\operatorname{Im} P = \left(\operatorname{Ker} P^*\right)^{\circ}.$$

Then the **index** of P is defined as

$$\operatorname{ind} P = \dim \operatorname{Ker} P - \operatorname{codim} \operatorname{Im} P = \dim \operatorname{Ker} P - \dim \operatorname{Ker} P^*.$$

1.8.2 Point Divisors

We will need to recall the rather technical notion of a **rigged divisor** from [31]. However, for reader's sake, we start with more familiar and easier to comprehend particular case of a **point divisor**, which appeared initially in Nadirashvili and Gromov-Shubin papers [30, 57].

Definition 1.21 A **point divisor** μ on X consists of two finite disjoint subsets of X

$$D^+ = \{x_1, \ldots, x_r\}, D^- = \{y_1, \ldots, y_s\} \tag{1.24}$$

and two tuplets $0 < p_1, \ldots, p_r$ and $q_1, \ldots, q_s < 0$ of integers. The **support of the point divisor** μ is $D^+ \bigcup D^-$. We will also write

$$\mu := x_1^{p_1} \ldots x_r^{p_r} \cdot y_1^{q_1} \ldots y_s^{q_s}.$$

In other words, μ is an element of the free abelian group generated by points of $\mathcal{X}$.

In [30, 57], such a divisor is used to allow solutions $u(x)$ of an elliptic equation $Pu = 0$ of order m on n-dimensional manifold X to have poles up to certain orders at the points of D^+ and enforce zeros on D^-. Namely,

(i) For any $1 \leq j \leq r$, there exists an open neighborhood U_j of x_j such that on $U_j \setminus \{x_j\}$, one has $u = u_s + u_r$, where $u_r \in C^\infty(U_j)$, $u_s \in C^\infty(U_j \setminus \{x_j\})$ and when $x \to x_j$,

$$u_s(x) = o(|x - x_j|^{m-n-p_j}).$$

(ii) For any $1 \leq j \leq s$, as $x \to y_j$, one has

$$u(x) = O(|x - y_j|^{|q_j|}).$$

1.8.3 Rigged Divisors

The notion of a "rigged" divisor comes from the desire to allow for some infinite sets $D^\pm$, but at the same time to impose only finitely many conditions ("zeros" and "singularities") on the solution.

So, let us take a deep breath and dive into it. First, let us define some distribution spaces:

Definition 1.22 For a closed set $C \subset \mathcal{X}$, we denote by $\mathcal{E}'_C(\mathcal{X})$ the space of distributions on $\mathcal{X}$, whose supports belong to C (i.e., they are zero outside C).

Definition 1.23

1. A **rigged divisor associated with** P is a tuple $\mu = (D^+, L^+; D^-, L^-)$, where $D^\pm$ are *compact nowhere dense disjoint subsets* in $\mathcal{X}$ and $L^\pm$ are *finite-dimensional* vector spaces of distributions on X supported in $D^\pm$ respectively, i.e.,

$$L^+ \subset \mathcal{E}'_{D+}(\mathcal{X}), \quad L^- \subset \mathcal{E}'_{D-}(\mathcal{X}).$$

2. The **secondary spaces** $\tilde{L}^\pm$ associated with $L^\pm$ are defined as follows:

$$\tilde{L}^+ = \{u \mid u \in \mathcal{E}'_{D+}(\mathcal{X}), \, Pu \in L^+\}, \quad \tilde{L}^- = \{u \mid u \in \mathcal{E}'_{D-}(\mathcal{X}), \, P^*u \in L^-\}.$$

3. Let $\ell^\pm = \dim L^\pm$ and $\tilde{\ell}^\pm = \dim \tilde{L}^\pm$. The **degree** of μ is defined as follows:

$$\deg_P \mu = (\ell^+ - \tilde{\ell}^+) - (\ell^- - \tilde{\ell}^-). \tag{1.25}$$

4. The **inverse of** μ is the rigged divisor $\mu^{-1} := (D^-, L^-; D^+, L^+)$ associated with P^*.

Remark 1.6

- Notice that the degree of the divisor involves the operator P, so it would have been more prudent to call it "degree of the divisor with respect to the operator P," but we'll neglect this, hoping that no confusion will arise.
- Observe that, due to their ellipticity, P and P^* are injective on $\mathcal{E}'_{D+}$ and $\mathcal{E}'_{D-}$, correspondingly.[10] Thus,

$$\ell^{\pm} \geq \tilde{\ell}^{\pm}. \tag{1.26}$$

- The sum of the degrees of a divisor μ and of its inverse is zero.

Although we have claimed that point divisors are also rigged divisors, this is not immediately clear when comparing the Definitions 1.21 and 1.23. Namely, we have to assign the spaces $L_{\pm}$ to a point divisor and to check that the definitions are equivalent in this case. This was done[11] in [31], if one defines the spaces associated with a point divisor as follows:

$$L^{+} = \left\{ \sum_{1 \leq j \leq r} \sum_{|\alpha| \leq p_j - 1} c_j^{\alpha} \delta^{\alpha}(\cdot - x_j) \mid c_j^{\alpha} \in \mathbb{C} \right\}$$

and

$$L^{-} = \left\{ \sum_{1 \leq j \leq s} \sum_{|\alpha| \leq |q_s| - 1} c_j^{\alpha} \delta^{\alpha}(\cdot - y_j) \mid c_j^{\alpha} \in \mathbb{C} \right\},$$

where δ and δ^{α} denote the Dirac delta function and its derivative corresponding to the multi-index α.

It was also shown in [31] that the degree $\deg_P(\mu)$ in this case is

$$\sum_{1 \leq j \leq r} \left[\binom{p_j + n - 1}{n} - \binom{p_j + n - 1 - m}{n} \right] \tag{1.27}$$

$$- \sum_{1 \leq j \leq s} \left[\binom{q_j + n - 1}{n} - \binom{q_j + n - 1 - m}{n} \right].$$

Here, as before, n is the dimension of the manifold X and m is the order of the operator P.

---

[10]For example, if $u \in \mathcal{E}'_{D+}$ and $Pu = 0$ then u is smooth due to elliptic regularity, but then $u = 0$ everywhere since the complement of D^+ is dense.

[11]Which is not trivial.

Remark 1.7 One observes a clear similarity between the combinatorial expressions in (1.27) and (1.18). It was one of the reasons to try to combine Liouville and Riemann-Roch type results.

1.8.4 Nadirashvili-Gromov-Shubin Theorem on Noncompact Manifolds

To state (a version of) the Gromov-Shubin theorem, we now introduce the spaces of solutions of P with allowed singularities on D^+ and vanishing conditions on D^-.

Notation 1.24 *For a compact subset K of X and $u \in C^\infty(X \setminus K)$, we shall write that*

$$u \in \mathrm{Dom}_K P,$$

if there is a compact neighborhood $\hat{K}$ of K and $\hat{u} \in \mathrm{Dom}\, P$ such that $u = \hat{u}$ outside $\overset{\circ}{\hat{K}}$.

Definition 1.25 For an elliptic operator P and a rigged divisor

$$\mu = (D^+, L^+; D^-, L^-),$$

the space $L(\mu, P)$ is defined as follows: $u \in L(\mu, P)$ iff $u \in \mathrm{Dom}_{D^+} P$ and there exists $\tilde{u} \in \mathcal{D}'(X)$, such that $\tilde{u} = u$ on $X \setminus D^+$, $P\tilde{u} \in L^+$, and $(u, L^-) = 0$.

Here $(u, L^-) = 0$ means that u is orthogonal to every element in L^- with respect to the canonical bilinear duality.

Remark 1.8 One notices that distributions $\tilde{u}$ are regularization of $u \in C^\infty(X \setminus D^+)$.

In other words, the space $L(\mu, P)$ consists of solutions of the equation $Pu = 0$ with poles allowed and zeros enforced by the divisor μ. It is worthwhile to notice that since the manifold is non-compact, the domain $\mathrm{Dom}\, P$ of the operator P will have to involve some conditions at infinity. This observation will be used later to treat a "pole" at infinity, i.e. Liouville property.

Now we can state a variant of Nadirashvili-Gromov-Shubin's version of the Riemann-Roch theorem.

Theorem 1.26 *Let P be an elliptic operator such that (1.19) and properties ($\mathcal{P}1$)–($\mathcal{P}4$) are satisfied. Let also μ be a rigged divisor associated with P. If P is a Fredholm operator on X, then the following Riemann-Roch inequality holds:*

$$\dim L(\mu, P) - \dim L(\mu^{-1}, P^*) \geq \mathrm{ind}\, P + \deg_P(\mu). \tag{1.28}$$

If both P and P* are Fredholm on $\mathcal{X}$, (1.28) becomes the Riemann-Roch equality:

$$\dim L(\mu, P) - \dim L(\mu^{-1}, P^*) = \text{ind } P + \deg_P(\mu). \tag{1.29}$$

Remark 1.9

1. Although the authors of [31] do not state their theorem in the exact form above, the Riemann-Roch inequality (1.28) follows from their proof.
2. If one considers the difference $\dim L(\mu, P) - \dim L(\mu^{-1}, P^*)$ as some "index of P in presence of the divisor μ" (say, denote it by $\text{ind}_\mu(P)$), the Riemann-Roch equality (1.29) becomes

$$\text{ind}_\mu(P) = \text{ind } P + \deg_P(\mu) \tag{1.30}$$

and thus it says that introduction of the divisor changes the index of the operator by $\deg_P(\mu)$.

Analogously, the inequality (1.28) becomes

$$\text{ind}_\mu(P) \geq \text{ind } P + \deg_P(\mu). \tag{1.31}$$

It is useful for our future considerations to mention briefly some of the ingredients[12] of the proof from [31]. To start, we define some auxiliary spaces. Let, as before, K be a nowhere dense compact set and we denote for a function $u \in C^\infty(X \setminus K)$ by $\tilde{u} \in \mathcal{D}'(\mathcal{X})$ its (non-uniquely defined) regularization. I.e., $u = \tilde{u}$ on $\mathcal{X} \setminus K$. Let

$$\Gamma(\mathcal{X}, \mu, P) := \{u \in C^\infty(X \setminus D^+) \mid u \in \text{Dom}_{D^+} P, \exists \tilde{u} \in \mathcal{D}'(\mathcal{X}) \text{ such that}$$

$$\tilde{u} = u \text{ on } \mathcal{X} \setminus D^+, \, P\tilde{u} \in L^+ + C^\infty(\mathcal{X}) \text{ and } \langle u, L^- \rangle = 0\}.$$

The regularization $\tilde{u}$ above is not unique, so we define the space of all such regularizations:

$$\tilde{\Gamma}(\mathcal{X}, \mu, P) := \{\tilde{u} \in \mathcal{D}'(\mathcal{X}) \mid \tilde{u}_{|\mathcal{X} \setminus D^+} \in \Gamma(\mathcal{X}, \mu, P), \, P\tilde{u} \in L^+ + C^\infty(\mathcal{X})\}.$$

It follows from the definition of the space $\Gamma(\mathcal{X}, \mu, P)$ that for any function u in this space, the function Pu (where P is applied as a differential expression) extends uniquely to a smooth function, which we call $\tilde{P}u$, on the whole $\mathcal{X}$. In the same manner, we can also define the corresponding extension $\tilde{P}^*$ as a linear map from $\Gamma(\mathcal{X}, \mu^{-1}, P^*)$ to $C^\infty(\mathcal{X})$.

---

[12]The reader interested in the main results only, can skip to Corollary 1.27.

Let us also introduce the spaces of functions "with enforced zeros":

$$\Gamma_\mu(\mathcal{X}, P) = \{u \in \mathrm{Dom}\, P \mid \langle u, L^- \rangle = 0\}$$

and

$$\tilde{\Gamma}_\mu(\mathcal{X}, P) = \{f \in \mathrm{Dom}'\, P^* \mid \langle f, \tilde{L}^- \rangle = 0\}.$$

An inspection of these definitions leads to the following conclusions:

Proposition 1.1

1. $\tilde{P}$ is a linear map from $\Gamma(\mathcal{X}, \mu, P)$ to $\tilde{\Gamma}_\mu(\mathcal{X}, P)$.
2. $\tilde{P}^*$ is a linear map from $\Gamma(\mathcal{X}, \mu^{-1}, P^*)$ to $\tilde{\Gamma}_{\mu^{-1}}(\mathcal{X}, P^*)$.
3. The spaces of solutions of interest are the kernels of the operators above:

$$L(\mu, P) = \mathrm{Ker}(\tilde{P}), \quad L(\mu^{-1}, P^*) = \mathrm{Ker}(\tilde{P}^*).$$

Let us also introduce the duality

$$(\cdot, \cdot) : \Gamma(\mathcal{X}, \mu, P) \times \tilde{\Gamma}_{\mu^{-1}}(\mathcal{X}, P^*) \to \mathbb{C} \tag{1.32}$$

as follows:

$$(u, f) := \langle \tilde{u}, f \rangle, \quad u \in \Gamma(\mathcal{X}, \mu, P), f \in \tilde{\Gamma}_{\mu^{-1}}(\mathcal{X}, P^*),$$

where $\tilde{u}$ is any element in the preimage of $\{u\}$ under the restriction map from $\tilde{\Gamma}(\mathcal{X}, \mu, P)$ to $\Gamma(\mathcal{X}, \mu, P)$. Similarly, we get the duality

$$(\cdot, \cdot) : \tilde{\Gamma}_\mu(\mathcal{X}, P) \times \Gamma(\mathcal{X}, \mu^{-1}, P^*) \to \mathbb{C} \tag{1.33}$$

These dualities are well-defined and non-degenerate [31]. Moreover, the following relation holds:

$$(\tilde{P}u, v) = (u, \tilde{P}^*v), \quad \forall u \in \Gamma(\mathcal{X}, \mu, P), v \in \Gamma(\mathcal{X}, \mu^{-1}, P^*).$$

by applying the additivity of Fredholm indices to some short exact sequences of the spaces introduced above (see [31, Lemma 3.1 and Remark 3.2]), Gromov and Shubin then establish the following basic facts:

Proposition 1.2 *[31, Lemma 3.1 – Lemma 3.4]*

1.

$$\dim \mathrm{Ker}\, \tilde{P} = \mathrm{ind}\, P + \deg_P(\mu) + \mathrm{codim}\, \mathrm{Im}\, \tilde{P}^*, \tag{1.34}$$

Note that the assumption that P is Fredholm on $\mathcal{X}$ is important for (1.34) to hold true.

2. $(\operatorname{Im} \tilde{P})^\circ = \operatorname{Ker} \tilde{P}^*$,
3. $\operatorname{Im} \tilde{P} \subset (\operatorname{Ker} \tilde{P}^*)^\circ$.
4.

$$\dim \operatorname{Ker} \tilde{P}^* = \operatorname{codim} (\operatorname{Ker} \tilde{P}^*)^\circ \leq \operatorname{codim} \operatorname{Im} \tilde{P}. \tag{1.35}$$

5. *The Riemann-Roch inequality (1.28) follows from (1.34) and (1.35).*
6. *If P^* is also Fredholm, one can apply (1.28) for P^* and μ^{-1} instead of P and μ to get*

$$\dim \operatorname{Ker} \tilde{P} \leq \operatorname{ind} P + \deg_P(\mu) + \dim \operatorname{Ker} \tilde{P}^*. \tag{1.36}$$

Now, the Riemann-Roch equality (1.29) follows from (1.28) and (1.36).

   In this case, as a byproduct of the proof of (1.29) [31, Theorem 2.12], one also gets $\operatorname{Im} \tilde{P} = (\operatorname{Ker} \tilde{P}^*)^\circ$ and $\operatorname{Im} \tilde{P}^* = (\operatorname{Ker} \tilde{P})^\circ$.

Here $(\operatorname{Im} \tilde{P})^\circ$ and $(\operatorname{Ker} \tilde{P}^*)^\circ$ are the annihilators of $\operatorname{Im} \tilde{P}$, $\operatorname{Ker} \tilde{P}^*$ with respect to the dualities (1.32) and (1.33), respectively.

Remark 1.10 If (1.35) becomes an equality, i.e.,

$$\dim \operatorname{Ker} \tilde{P}^* = \operatorname{codim} \operatorname{Im} \tilde{P},$$

one obtains the Riemann-Roch equality (1.29) for the rigged divisor μ without assuming that P^* is Fredholm on $\mathcal{X}$. Conversely, if (1.29) holds, then $\operatorname{Im} \tilde{P} = (\operatorname{Ker} \tilde{P}^*)^\circ$.

As a result, we have the following useful corollary:

Corollary 1.27 *Let P be Fredholm on $\mathcal{X}$, $\operatorname{Im} P = \operatorname{Dom}' P^*$, and*

$$\mu = (D^+, L^+; D^-, L^-)$$

be a rigged divisor on $\mathcal{X}$. Then the Riemann-Roch equality (1.29) holds for P and divisor μ.

   Moreover, the space $L(\mu^{-1}, P^*)$ is trivial, if the following additional condition is satisfied: If u is a smooth function in $\operatorname{Dom} P$ such that $\langle Pu, \tilde{L}^- \rangle = 0$, then there exists a solution v in $\operatorname{Dom} P$ of the equation $Pv = 0$ satisfying $\langle u - v, L^- \rangle = 0$.

   In particular, this assumption holds automatically if $D^- = \emptyset$.

We end this section by recalling an application of Theorem 1.26, which will be used later.

Example 1.1 ([31, Example 4.6] and [57]) Consider $P = P^* = -\Delta$ on $\mathcal{X} = \mathbb{R}^d$, where $d \geq 3$ and

$$\text{Dom } P = \text{Dom } P^* = \{u \mid u \in C^\infty(\mathbb{R}^d),\ \Delta u \in C_c^\infty(\mathbb{R}^d) \text{ and } \lim_{|x|\to\infty} u(x) = 0\},$$

$$\text{Dom}' P = \text{Dom}' P^* = C_c^\infty(\mathbb{R}^d).$$

Then the operators P and P^* are Fredholm on $\mathbb{R}^d$, Ker $P = $ Ker $P^* = \{0\}$, Im $P = $ Im $P^* = C_c^\infty(\mathbb{R}^d)$, and thus ind $P = 0$ (see [31, Example 4.2]).

Let

$$D^+ = \{y_1, \ldots, y_k\}, \quad D^- = \{z_1, \ldots, z_l\}.$$

with all the points $y_1, \ldots, y_k, z_1, \ldots, z_l$ pairwise distinct. Consider the following distributional spaces: L^+ is the vector space spanned by Dirac delta distributions $\delta(\cdot - y_j)$ supported at the points y_j ($1 \leq j \leq k$); L^- is spanned by the first order derivatives $\dfrac{\partial}{\partial x_\alpha}\delta(\cdot - z_j)$ of Dirac delta distributions supported at z_j ($1 \leq j \leq l$, $1 \leq \alpha \leq d$).[13]

Consider now the rigged divisor $\mu := (D^+, L^+; D^-, L^-)$. Then $\deg_{-\Delta}(\mu) = k - dl$. Furthermore,

$$L(\mu, -\Delta)$$
$$= \left\{u \mid u(x) = \sum_{j=1}^k \frac{a_j}{|x - y_j|^{d-2}},\, a_j \in \mathbb{C}, \text{ and } \nabla u(z_j) = 0,\, j = 1, \ldots, l\right\},$$

and

$$L(\mu^{-1}, -\Delta) = \left\{v \mid v(x) = \sum_{j=1}^l \sum_{\alpha=1}^d b_{j,\alpha} \frac{\partial}{\partial x_\alpha}\left(|x - z_j|^{2-d}\right),\right.$$

$$\left. b_{j,\alpha} \in \mathbb{C}, \text{ and } u(y_j) = 0,\, j = 1, \ldots, k\right\}.$$

In this case, the Nadirashvili-Gromov-Shubin Riemann-Roch-type formula (Theorem 1.26) is

$$\dim L(\mu, -\Delta) - \dim L(\mu^{-1}, -\Delta) = k - dl. \tag{1.37}$$

---

[13]Note that the secondary spaces $\tilde{L}^\pm$ are trivial.

Chapter 2
The Main Results

Abstract This chapter introduces the objects required for formulating the main results of Lioville-Riemann-Roch type. The results are formulated. In some cases, the Riemann-Roch type equalities cannot be achieved (counterexamples are shown), while inequalities still hold. These inequalities, however, can be applied, the same way the equalities are, for proving the existence of solutions of elliptic equations with prescribed zeros, poles, and growth at infinity.

In this chapter, we consider a periodic elliptic operator A of order m on an n-dimensional co-compact abelian covering

$$X \underset{\mathbb{Z}^d}{\mapsto} M.$$

Notice, again, that the rank d of the deck group does not have to be related in any way to the dimension n of the manifold.

To make a combination of Liouville and Riemann-Roch theorems meaningful, we assume that the Liouville property holds for the operator A at the level $\lambda = 0$, i.e. (see Theorem 1.11), the **real** Fermi surface of A (see Definition 1.5) is finite (modulo G^*-shifts).

The approach we will follow consists in finding appropriate functional spaces that would incorporate the polynomial growth at infinity and that could be handled by the general techniques and results of Gromov and Shubin.

There are two significantly different possibilities: (1) The Fermi surface finite, but non-empty. (2) The Fermi surface is empty. We start with the more interesting first one.

2.1 Non-empty Fermi Surface

Suppose that $F_{A,\mathbb{R}} = \{k_1, \dots, k_\ell\}$ (modulo G^*-shifts), where $\ell \geq 1$.

© The Author(s), under exclusive license to Springer Nature Switzerland AG 2021   23
M. Kha, P. Kuchment, *Liouville-Riemann-Roch Theorems on Abelian Coverings*,
Lecture Notes in Mathematics 2245, https://doi.org/10.1007/978-3-030-67428-1_2

2.1.1 Assumptions

We need to make the following assumption on the local behavior of the Bloch variety of the operator A around each quasimomentum k_j in the real Fermi surface:

Assumption $\mathcal{A}$

($\mathcal{A}$1) *For any quasimomentum k, the spectrum of the operator $A(k)$ is discrete.*

Under this assumption, the following lemma can be deduced immediately from Proposition 5.3 and perturbation theory (see e.g., [33, 61]):

Lemma 2.1 *For each quasimomentum $k_r \in F_{A,\mathbb{R}}$, there is an open neighborhood V_r of k_r in $\mathbb{R}^d$ and a closed contour $\Upsilon_r \subset \mathbb{C}$, such that*

a. *The neighborhoods V_r are mutually disjoint;*
b. *The contour Υ_r surrounds the eigenvalue 0 and does not contain any other points of the spectrum $\sigma(A(k_r))$;*
c. *The intersection $\sigma(A(k)) \cap \Upsilon_r$ is empty for any $k \in V_r$.*

Then, for any $k \in V_r$, we can define the Riesz projector

$$\Pi_r(k) := \frac{1}{2\pi i} \oint_{\Upsilon_r} (A(k) - zI)^{-1} dz$$

associated with $A(k)$ and the contour Υ_r. Thus, $\Pi_r(k)A(k)$ is well-defined for any $k \in V_r$. Let m_r be the algebraic multiplicity of the eigenvalue 0 of the operator $A(k_r)$. The immediate consequence is:

Lemma 2.2 *The projector $\Pi_r(k)$ depends analytically on $k \in V_r$. In particular, its range $R(\Pi_r(k))$ has the same dimension m_r for all $k \in V_r$ and the union $\bigcup_{k \in V_r} R(\Pi_r(k))$ forms a trivial holomorphic vector bundle over V_r.*

We denote by $A_r(k)$ the matrix representation of the operator

$$\Pi_r(k)A(k)|_{R(\Pi_r(k))}$$

with respect to a fixed holomorphic basis $(f_j(k))_{1 \leq j \leq m_r}$ of the range $R(\Pi_r(k))$ when $k \in V_r$. Then $A_r(k)$ is an invertible matrix except only for $k = k_r$. We equip $\mathbb{C}^{m_r}$ with the maximum norm and impose the following integrability condition:

($\mathcal{A}$2)

$$\int_{V_r \setminus \{k_r\}} \|A_r(k)^{-1}\|_{\mathcal{L}(\mathbb{C}^{m_r})} dk < \infty, \quad \text{for all } r = 1, \dots, \ell,$$

where $\mathcal{L}(\mathbb{C}^{m_r})$ is the algebra of linear operators on $\mathbb{C}^{m_r}$.

Remark 2.1

(i) Thanks to Proposition 5.2, Assumption $(\mathcal{A}1)$ is satisfied if A is either self-adjoint or a real operator of even order.[1]

(ii) When the rank d of G is greater than 2, Assumption $(\mathcal{A}2)$ holds at a **generic** spectral edge (see Chap. 4 and [34, 37, 39, 42–44, 46]).

2.1.2 Spaces

To formulate the results and to be able to use the conclusions of Gromov and Shubin, it is crucial to define spaces of solutions of the equation $Au = 0$ that combine polynomial growth at infinity with satisfying the conditions imposed by a rigged divisor μ.

Definition 2.1 Given any $p \in [1, \infty]$ and $N \in \mathbb{R}$, we define

$$L_p(\mu, A, N) := L(\mu, A_N^p),$$

where the operator A_N^p stands for A with the domain

$$\operatorname{Dom} A_N^p = \{u \in V_N^p(X) \mid Au \in C_c^\infty(X)\}.$$

In other words, $L_p(\mu, A, N)$ is the space

$$\{u \in \operatorname{Dom}_{D^+} A_N^p \mid \exists \tilde{u} \in \mathcal{D}'(X) : \tilde{u} = u \text{ on } X \setminus D^+, A\tilde{u} \in L^+, (u, L^-) = 0\}.$$

We thus restrict the growth (in L_p-sense) of function u at infinity to a polynomial of order N, impose the divisor μ conditions, and require that u satisfies the homogeneous equation $Au = 0$ outside the compact D^+.

Remark 2.2 Consider $u \in L_p(\mu, A, N)$. Let K be a compact domain in X such that $\bigcup_{g \in G} gK = X$. Define $G_{K,D^+} := \{g \in G \mid \operatorname{dist}(gK, D^+) \geq 1\}$, where we use the notation $\operatorname{dist}(\cdot, \cdot)$ for the distance between subsets arising from the Riemannian distance on X. Since $Au = 0$ on $X \setminus D^+$, the condition "$u \in \operatorname{Dom}_{D^+} A_N^p$" can be written equivalently as follows:

$$\{\|u\|_{L^2(gK)} \cdot \langle g \rangle^{-N}\}_{g \in G_{K,D^+}} \in \ell^p(G_{K,D^+}).$$

---

[1] Here A is real means that Au is real whenever u is real.

By Schauder estimates (see Proposition 5.5), this condition can be rephrased as follows:

$$\sup_{x:\,\mathrm{dist}(x,D^+)\geq 1} \frac{|u(x)|}{\mathrm{dist}(x,D^+)^N} < \infty, \qquad\qquad \text{when} \quad p = \infty,$$

$$\int_{x:\,\mathrm{dist}(x,D^+)\geq 1} \frac{|u(x)|^p}{\mathrm{dist}(x,D^+)^{pN}}\mathrm{d}\mu_X(x) < \infty, \quad \text{when} \quad 1 \leq p < \infty. \tag{2.1}$$

So, depending on the sign of N, this condition controls how u grows or decays at infinity.

2.1.3   Results

Our first main result is contained in the next theorem, establishing a Liouville-Riemann-Roch type inequality.

Theorem 2.2 *Assume that either $p = \infty$ and $N \geq 0$ or $p \in [1, \infty)$ and $N > d/p$. Let p' be the Hölder conjugate of p. Then, under Assumption $\mathcal{A}$ imposed on the operator A, the following Liouville-Riemann-Roch inequality holds:*

$$\dim L_p(\mu, A, N) - \dim L_{p'}(\mu^{-1}, A^*, -N) \geq \dim V_N^p(A) + \deg_A(\mu), \tag{2.2}$$

where $\dim V_N^p(A)$ can be computed via Theorems 1.17 and 1.13.

Remark 2.3

- This is an extension of the Riemann-Roch inequality (1.28) to include also Liouville conditions on growth at infinity.
- One may wonder why in comparison with the "μ-index" $\dim L_p(\mu, A, N) - \dim L_p(\mu^{-1}, A^*, -N)$, the above inequality only involves the **dimension of the kernel** $V_N^p(A)$, rather than a full index. The reason is that in this case, the two coincide, the co-dimension of the range being equal to zero.
- Assumption $(\mathcal{A}1)$ forces the Fredholm index of $A(0)$ on M to vanish (see [41, Theorem 4.1.4]). Therefore (by Atiyah's theorem [5]), $\mathrm{ind}_M A(k)$ ($k \in \mathbb{C}^d$) and the L^2-index of A are equal to zero as well.

As an useful direct corollary of Theorem 2.2, we obtain

Theorem 2.3 *If $\dim V_N^p(A) + \deg_A(\mu) > 0$, then there exists a nonzero element in the space $L_p(\mu, A, N)$.*

In other words, there exists a nontrivial solution of $Au = 0$ of the growth described in (2.1) that also satisfies the conditions on "zeros" and "singularities" imposed by the rigged divisor μ.

To state the next results, we will need the following definition that addresses divisors containing only "zeros" or "singularities".

Definition 2.4

- The divisor $(\emptyset, \{0\}; \emptyset, \{0\})$ is called **trivial**.
- Let $\mu = (D^+, L^+; D^-, L^-)$ be a rigged divisor on X. Then its **positive and negative parts** are defined as follows:

$$\mu^+ := (D^+, L^+; \emptyset, \{0\}), \quad \mu^- := (\emptyset, \{0\}; D^-, L^-).$$

E.g., for a rigged divisor μ, μ^+ (resp. μ^-) is trivial whenever $D^+ = \emptyset$ (respectively, $D^- = \emptyset$).

Our next result shows that if μ is positive (i.e., μ^- is trivial), the Liouville-Riemann-Roch inequality becomes an equality:

Theorem 2.5 *Let μ be a rigged divisor and μ^+ - its positive part. Under the assumptions of Theorem 2.2,*

1. The space $L_{p'}((\mu^+)^{-1}, A^*, -N)$ is trivial (i.e., contains only zero).
2.

$$\dim L_p(\mu^+, A, N) = \dim V_N^p(A) + \deg_A(\mu^+). \tag{2.3}$$

3. In particular,

$$\dim L_p(\mu, A, N) \le \dim V_N^p(A) + \deg_A(\mu^+). \tag{2.4}$$

In other words, the inequality (2.4) gives an upper bound for the dimension of the space $L_p(\mu, A, N)$ (see (2.2) for its lower bound) in terms of the degree of the positive part of the divisor μ and the dimension of the space $V_N^p(A) = L_p(\mu_0, A, N)$, where μ_0 is the trivial divisor. Then (2.3) shows that this estimate saturates for divisors containing only "singularities".

When μ^- is nontrivial, determining the triviality of the space

$$L_{p'}(\mu^{-1}, A^*, -N)$$

is more complicated. In the next proposition we show that if the degree of μ^+ is sufficiently large, the space $L_{p'}(\mu^{-1}, A^*, -N)$ degenerates to zero, while the spaces $L_{p'}(\mu^{-1}, A^*, -N)$ can have arbitrarily large dimensions if μ^+ is trivial.

We need to recall the following definition (e.g., [54]):

Definition 2.6 A differential operator A is said to have the **strong unique continuation property** if any local solution of the equation $Au = 0$ that vanishes to the infinite order at a point, vanishes identically.

We can now state the promised proposition:

Proposition 2.7

(a) *For any $N \geq 0$, $p \in [1, \infty]$, and $d \geq 3$,*

$$\sup_{\mu=\mu^-} \dim L_{p'}(\mu^{-1}, -\Delta_{\mathbb{R}^d}, -N) = \infty,$$

where the supremum is taken over all divisors μ with trivial positive parts.

(b) *Let the assumptions of Theorem 2.2 be satisfied and A^* have the strong unique continuation property. Let also the covering X be connected and a point x_0 of X selected. Suppose that $p \in [1, \infty]$, $N \in \mathbb{R}$, a compact nowhere dense set $D^- \subset X \setminus \{x_0\}$, and a finite dimensional subspace L^- of $\mathcal{E}'_{D^-}(X)$ are fixed. Then there exists $M > 0$ depending on (A, p, N, x_0, D^-, L^-), such that*

$$\dim L_{p'}(\mu^{-1}, A^*, -N) = 0$$

for any rigged divisor $\mu = (D^+, L^+; D^-, L^-)$ such that $x_0 \in D^+ \subseteq X \setminus D^-$ and L^+ contains the linear span

$$\mathcal{L}_M^+ := \operatorname{span}_{\mathbb{C}}\{\partial_x^\alpha \delta(\cdot - x_0)\}_{0 \leq |\alpha| \leq M}.$$

The second part of the above proposition is a reformulation of [30, Proposition 4.3].

Theorem 2.2 can be improved under the following **strengthened version of** Assumption $\mathcal{A}$:

For each $1 \leq r \leq \ell$, the function $k \in V_r \mapsto \|A_r(k)^{-1}\|^2_{\mathcal{L}(\mathbb{C}^{m_r})}$ is integrable. (2.5)

This happens, for instance, at generic spectral edges if $d \geq 5$.

Theorem 2.8 *Let the assumption (2.5) above be satisfied. Then, if*

- *either $p \geq 2$ and $N \geq 0$,*
- *or $p \in [1, 2)$ and $2pN > (2 - p)d$,*

then the inequality (2.2) holds.

In particular, for any rigged divisor μ,

$$\dim L_2(\mu^+, A, 0) = \deg_A(\mu^+) \tag{2.6}$$

and

$$\dim L_2(\mu, A, 0) = \deg_A(\mu) + \dim L_2(\mu^{-1}, A^*, 0). \tag{2.7}$$

Remark 2.4 When μ is trivial, the equality (2.6) means the absence of non-zero L^2- solutions (bound states). Thus, generically, spectral edges are not eigenvalues. In this case, the condition on integrability of $\|A_r(k)^{-1}\|^2_{\mathcal{L}(\mathbb{C}^{m_r})}$ is not required.

The following L^2-solvability result (or Fredholm alternative) follows from (2.7).

Proposition 2.9 *Let D be a non-empty compact nowhere dense subset of X and a finite dimensional subspace $L \subset \mathcal{E}'_D(X)$ be given. Define the finite dimensional subspace $\tilde{L} = \{v \in \mathcal{E}'_D(X) \mid A^*v \in L\}$. Consider any $f \in C_c^\infty(X)$ satisfying $(f, \tilde{L}) = 0$. Under the assumptions of Theorem 2.8 for the periodic operator A of order m, the following two statements are equivalent:*

 (i) *f is orthogonal to each element in the space $L_2(\mu, A^*, 0)$, where μ is the rigged divisor $(D, L; 0, \emptyset)$.*
 (ii) *The inhomogeneous equation $Au = f$ has a (unique) solution u in $H^m(X)$ such that $(u, L) = 0$.*

We describe now examples that show that when μ is negative (i.e., μ^+ is trivial), the Liouville-Riemann-Roch equality might fail miserably (while still holding for some negative divisors).

Proposition 2.10 *Consider the Laplacian $-\Delta$ on $\mathbb{R}^d$, $d \geq 3$. For any $N \geq 0$ and positive integer ℓ, there exists a rigged divisor $\mu = \mu^-$, such that*

$$\dim L_\infty(\mu, -\Delta, N) - \dim L_1(\mu^{-1}, -\Delta, -N) \geq \ell + \dim V_N^\infty(-\Delta) + \deg_{-\Delta}(\mu).$$
(2.8)

In other words, the difference between left and right hand sides of the Liouville-Riemann-Roch inequality can be made arbitrarily large.

On the other hand, one can also achieve an equality for a negative divisor $\mu = \mu^-$. Indeed, let $A = A(D)$ be an elliptic constant-coefficient homogeneous differential operator of order m on $\mathbb{R}^d$ that satisfies Assumption $\mathcal{A}$. Consider two non-negative integers $M_0 \geq M_1$, fix a point x_0 in X, and define $D^- := \{x_0\}$ and the following finite dimensional vector subspace of $\mathcal{E}'_{D^-}(\mathbb{R}^d)$:

$$L^- := \text{span}_{\mathbb{C}}\{\partial^\alpha \delta(\cdot - x_0)\}_{M_1 \leq |\alpha| \leq M_0}.$$

Let μ be a rigged divisor on $\mathbb{R}^d$ of the form $(D^+, L^+; D^-, L^-)$, where D^+ is a nowhere dense compact subset of $\mathbb{R}^d \setminus \{x_0\}$ and L^+ is a finite dimensional subspace of $\mathcal{E}'_{D^+}(\mathbb{R}^d)$.[2]

---

[2] A particular case is when $M_1 = 0$ and μ^+ is trivial, i.e., $\mu = \mu^-$ becomes the point divisor $x_0^{-(M_0+1)}$ (see Definition 1.21).

Proposition 2.11 *Assume that one of the following two conditions holds:*

- *either* $1 \leq p < \infty$, $N > d/p + M_0$,
- *or* $p = \infty$, $N \geq M_0$.

Then $\dim L_{p'}(\mu^{-1}, A^*, -N) = 0$ *and the Liouville-Riemann-Roch equality holds:*

$$\dim L_p(\mu, A, N) = \dim V_N^p(A) + \deg_A(\mu).$$

Corollary 2.12 *Suppose that* $\{A_z\}_z$ *and* $\{\mu_z\}_z$ *are families of such periodic elliptic operators* A_z *satisfying the same assumption of Theorem 2.8 and of rigged divisors* μ_z *that depend continuously on a parameter* z. *Then the functions* $z \mapsto \dim L_2(\mu_z, A_z, 0)$ *and* $z \mapsto \dim L_2(\mu_z^{-1}, A_z^*, 0)$ *are upper-semicontinuous.*

2.2 Empty Fermi Surface

For a periodic elliptic operator A, the claim of emptiness of the (real) Fermi surface (as always in this text, at the $\lambda = 0$ level) is equivalent to its spectrum not containing zero (see Theorem 1.4):

$$0 \notin \sigma(A).$$

Interpreting the Fermi surface emptiness condition this way allows us to apply it even for non-periodic operators, the opportunity that we will use in this section.

Let us look at the periodic case first. Then the Liouville theorem becomes trivial, because the emptiness of the Fermi surface and thus zero being not in the spectrum implies that there is no non-zero polynomially growing solution (see [44, Theorem 4.3], which is an analog of the Schnol' theorem, see e.g., [19, 26, 69]).[3] As we show below, one can obtain now a Liouville-Riemann-Roch type result by combining the Riemann-Roch and the Schnol theorems. In fact, this will be done in the much more general case of C^∞-bounded uniformly elliptic operators (not necessarily periodic) on a co-compact (not necessarily abelian) Riemannian covering of sub-exponential growth. We will use here the results of [66] showing that a more general and stronger statement than the Schnol theorem holds for any C^∞-bounded uniformly elliptic operator on a manifold of bounded geometry and sub-exponential growth.

We begin with some definitions from [66] first.[4] Let $\mathcal{X}$ be a co-compact connected Riemannian covering and $\mathcal{M}$ be its base. The deck group G is a

---

[3]The Schnol' type theorems claim that under appropriate conditions presence of a non-trivial solution of sub-exponential growth implies presence of spectrum. The example of hyperbolic plane shows that this is not always true, but a correct re-formulation of the sub-exponential growth condition [13, Theorem 3.2.2 for the quantum graph case and Section 3.8] fixes this issue.

[4]One could find details about analysis on manifolds of bounded geometry in e.g., [21, 66, 67].

countable, finitely generated, and discrete (not necessarily abelian). Let $d_{\mathcal{X}}(\cdot, \cdot)$ be the G-invariant Riemannian distance on $\mathcal{X}$. Due to the compactness of $\mathcal{M}$, there exists $r_{inj} > 0$ (injectivity radius) such that for every $r \in (0, r_{inj})$ and every $x \in \mathcal{X}$, the exponential geodesic mapping $\exp_x : T\mathcal{X}_x \to \mathcal{X}$ is a diffeomorphism of the Euclidean ball $B(0, r)$ centered at 0 with radius r in the tangent space $T\mathcal{X}_x$ onto the geodesic ball $B_{\mathcal{X}}(x, r)$ centered at x with the same radius r in $\mathcal{X}$. Taking $r_0 \in (0, r_{inj})$, the geodesic balls $B_{\mathcal{X}}(x, r)$, where $0 < r \le r_0$, are called **canonical charts** with x-coordinate in the charts.

Definition 2.13 ([66, 67])

(i) A differential operator P of order m on $\mathcal{X}$ is C^∞-**bounded** if in every canonical chart, P can be represent as $\sum_{|\alpha| \le m} a_\alpha(x)\partial_x^\alpha$, where the coefficients $a_\alpha(x)$ are smooth and for any multi-index β, $|\partial_x^\beta a_\alpha(x)| \le C_{\alpha\beta}$, where the constants $C_{\alpha\beta}$ are independent of the chosen canonical coordinates.

(ii) A differential operator P of order m on $\mathcal{X}$ is **uniformly elliptic**[5] if

$$|P_0^{-1}(x, \xi)| \le C|\xi|^{-m}, \quad (x, \xi) \in T^*\mathcal{X}, \xi \ne 0.$$

Here $T^*\mathcal{X}$ is the cotangent bundle of $\mathcal{X}$, $P_0(x, \xi)$ is the principal symbol of the operator P, and $|\xi|$ is the length of the covector (x, ξ) with respect to the metric on $T^*\mathcal{X}$ induced by the Riemannian metric on $\mathcal{X}$.

(iii) $\mathcal{X}$ is of **subexponential growth** if the volumes of balls of radius r grow subexponetially as $r \to \infty$, i.e., for any $\epsilon > 0$ and $r > 0$,

$$\sup_{x \in \mathcal{X}} \text{vol } B(x, r) = O(\exp(\epsilon r)).$$

Here $\text{vol}(\cdot)$ is the Riemannian volume on $\mathcal{X}$.

(iv) Let x_0 be a fixed point in $\mathcal{X}$. A differential operator P on $\mathcal{X}$ satisfies **Strong Schnol Property** (SSP) if the following statement is true: If there exists a non-zero solution u of the equation $Pu = \lambda u$ such that for any $\epsilon > 0$

$$u(x) = O(\exp(\epsilon d_{\mathcal{X}}(x, x_0)))$$

then λ is in the spectrum of P.

We now turn to a brief discussion of growth of groups (see e.g., [27, 28, 58]). Let us pick a finite, symmetric generating set S of G. The **word metric** associated to S is denoted by $d_S : G \times G \to \mathbb{R}$, i.e., for every pair (g_1, g_2) of two group elements in G, $d_S(g_1, g_2)$ is the length of the shortest representation in S of $g_1^{-1} g_2$ as a product of generators from S. Let e be the identity element of G. The volume function of G associated to S is the function $\text{vol}_{G,S} : \mathbb{N} \to \mathbb{N}$ defined by assigning to every $n \in \mathbb{N}$

---

[5]Clearly, any G-periodic elliptic differential operator with smooth coefficients on $\mathcal{X}$ is C^∞-bounded uniformly elliptic.

the cardinality of the open ball $B_{G,S}(e, n)$ centered at e with radius n in the metric space (G, d_S). Although the values of this volume function depend on the choice of the generating set S, its asymptotic growth type is independent of it. The group G is said to be of **subexponential growth** if

$$\lim_{n \to \infty} \frac{\ln \text{vol}_{G,S}(n)}{n} = 0.$$

It is known that the deck group G is of subexponential growth if and only if the covering $\mathcal{X}$ is so (see e.g., [64, Proposition 2.1]). Virtually nilpotent groups clearly have polynomial growth, and the celebrated Gromov's theorem [28] shows that they are the only ones. Thus, any virtually nilpotent co-compact Riemannian covering $\mathcal{X}$ is of subexponential growth. Groups with intermediate growth, which were constructed by Grigorchuk [27], provide other non-trivial examples of Riemannian coverings with subexponential growth.

Theorem 2.14 ([66, Theorem 4.2]) *If $\mathcal{X}$ is of subexponential growth, then any C^∞-bounded uniformly elliptic operator on $\mathcal{X}$ satisfies (SSP).*

Remark 2.5 A Schnol type theorem can be established without the subexponential growth condition, if the growth of a generalized eigenfunction is controlled in an integral (over an expanding ball), rather than point-wise sense. See [13, Theorem 3.2.2 for the quantum graph case and Section 3.8].

Similarly, we also say that a positive function $\varphi : G \to \mathbb{R}^+$ is of **subexponential growth** if

$$\lim_{|g| \to \infty} \frac{\ln \varphi(g)}{|g|} = 0,$$

where the **word length** of $g \in G$ is defined as $|g| := d_S(e, g)$. Again, this concept does not depend on the choice of the finite generating set S (see [58, Theorem 1.3.12]).

Definition 2.15 Let φ be a positive function defined on the deck group G such that both φ and its inverse φ^{-1} are of subexponential growth. Let us denote by $\mathcal{S}(G)$ the set of all such φ on G. Then for any $p \in [1, \infty]$ and $\varphi \in \mathcal{S}(G)$, we define:

$$\mathcal{V}_\varphi^p(\mathcal{X}) = \{u \in C^\infty(\mathcal{X}) \mid \{\|u\|_{L^2(g\mathcal{F})} \varphi^{-1}(g)\}_{g \in G} \in \ell^p(G)\},$$

where $\mathcal{F}$ is a fundamental domain for $\mathcal{M}$ in $\mathcal{X}$.

Also, let P_φ^p be the operator P with the domain $\{u \in \mathcal{V}_\varphi^p(\mathcal{X}) \mid Pu \in C_c^\infty(\mathcal{X})\}$. We denote by $L_p(\mu, P, \varphi)$ the space $L(\mu, P_\varphi^p)$, where μ is a rigged divisor on $\mathcal{X}$. In a similar manner, we also define the space $L_p(\mu, P^*, \varphi)$, where P^* is the transpose of P. In particular, if $G = \mathbb{Z}^d$ and $\varphi(g) = \langle g \rangle^N$, $\mathcal{V}_\varphi^p(\mathcal{X})$ is the space $V_N^p(\mathcal{X})$ introduced in Definition 1.9, while $L_p(\mu, P, \varphi)$ coincides with the space $L_p(\mu, P, N)$ appearing in Definition 2.1.

We can now state our result.

Theorem 2.16 *Consider any Riemannian co-compact covering $\mathcal{X}$ of subexponential growth with a discrete deck group G. Let P be a C^∞-bounded uniformly elliptic differential operator P of order m on $\mathcal{X}$ such that $0 \notin \sigma(P)$. Let us denote by φ_0 the constant function 1 defined on G. Then the following statements are true:*

(a) *For each rigged divisor μ on $\mathcal{X}$, $L_p(\mu, P, \varphi) = L_\infty(\mu, P, \varphi_0)$, where $p \in [1, \infty]$ and $\varphi \in \mathcal{S}(G)$. Thus, all the spaces $L_p(\mu, P, \varphi)$ are the same.*

(b) $\dim L_\infty(\mu, P, \varphi_0) = \deg_P(\mu) + \dim L_\infty(\mu^{-1}, P^*, \varphi_0)$.

(c) *If μ is a positive divisor, i.e. $\mu = (D^+, L^+; \emptyset, 0)$, then $\dim L_\infty(\mu, P, \varphi_0) = \deg_P(\mu)$.*

Now let $D \subset \mathcal{X}$ be a compact nowhere dense subset, L be a finite dimensional subspace of $\mathcal{E}'_D(\mathcal{X})$, and $\mu := (D, L; 0, \emptyset)$ be a positive divisor.

We define the space

$$\tilde{L} := \{u \in \mathcal{E}'_D(\mathcal{X}) \mid P^* u \in L\}.$$

The following analog of Corollary 2.9 holds:

Corollary 2.17 *Let the assumptions of Theorem 2.16 hold and a function $f \in C_c^\infty(X)$ be given such that $(f, \tilde{L}) = 0$. The following statements are equivalent:*

(i) *f is orthogonal to the vector space $L_\infty(\mu, P^*, \varphi_0)$.*

(ii) *There exists a unique solution u of the inhomogeneous equation $Pu = f$ such that $(u, L) = 0$ and $u \in \mathcal{V}_\varphi^p(\mathcal{X})$ for some $p \in [1, \infty]$ and $\varphi \in \mathcal{S}(G)$.*

(iii) *The equation $Pu = f$ admits a unique solution u which has subexponential decay and satisfies $(u, L) = 0$.*

(iv) *The equation $Pu = f$ admits a unique solution u which has exponential decay and satisfies $(u, L) = 0$.*

Remark 2.6

(i) Comparing to the Riemann-Roch formula (1.29), the Fredholm index of P does not appear in the formula in Theorem 2.16 (b) since P is invertible in this case.

(ii) When μ is trivial, Theorem 2.16 (c) becomes Theorem 2.14 and our Corollary 2.17 is an analog of [41, Theorem 4.2.1] for the co-compact Riemannian coverings of subexponential growth.

Chapter 3
Proofs of the Main Results

Abstract Here the proofs of the main results are provided, modulo some more technical auxiliary parts, which are delegated to Chap. 5.

3.1 Some Notions

First, we introduce some notions.

Definition 3.1 We will often use the notation $A \lesssim B$ to indicate that the quantity A is less or equal than the quantity B up to some multiplicative constant factor, which does not affect the arguments.

Definition 3.2 For each $s, N \in \mathbb{R}$ and $p \in [1, \infty]$, we denote by $V_{s,N}^p(X)$ the vector space consisting of all function $u \in C^\infty(X)$ such that for some (and thus any) compact subset K of X satisfying (1.13), the sequence $\{\|u\|_{H^s(gK)}\langle g\rangle^{-N}\}_{g \in G}$ belongs to $\ell^p(G)$. For a periodic elliptic operator A we put

$$V_{s,N}^p(A) := V_{s,N}^p(X) \cap \text{Ker } A.$$

Let also $A_{s,N}^p$ be the elliptic operator A with the domain

$$\text{Dom } A_{s,N}^p = \{u \in V_{s,N}^p(X) \mid Au \in C_c^\infty(X)\}.$$

When $s = 0$, this reduces to the notions of $V_N^p(X)$ and $V_N^p(A)$ introduced in Definition 1.9, and of the operator A_N^p and its domain Dom A_N^p in Definition 2.1.

3.2 Proof of Theorem 2.2

So, let A be a periodic elliptic differential operator of order m on X and a pair (p, N) satisfy the assumption of the theorem.

M. Kha, P. Kuchment, *Liouville-Riemann-Roch Theorems on Abelian Coverings*, Lecture Notes in Mathematics 2245, https://doi.org/10.1007/978-3-030-67428-1_3

Let $\mathcal{F}$ be the closure of a fundamental domain for G-action on X. We also pick a compact neighborhood $\hat{\mathcal{F}}$ in X of $\mathcal{F}$, so the conclusion of Proposition 5.5 applies.

Our proof will be done in several steps.

Step 1 We claim that given $p \in [1, \infty]$, $N \in \mathbb{R}$, and any rigged divisor $\mu = (D^+, L^+; D^-, L^-)$, one has

$$L(\mu, A^p_{m,N}) = L(\mu, A^p_N) = L_p(\mu, A, N). \tag{3.1}$$

Indeed, it suffices to show that $L(\mu, A^p_N) \subseteq L(\mu, A^p_{m,N})$.

Consider $u \in L(\mu, A^p_N)$. Due to Remark 2.2, this implies that

$$\{\|u\|_{L^2(g\hat{\mathcal{F}})} \cdot \langle g \rangle^{-N}\}_{G_{\hat{\mathcal{F}}, D^+}} \in \ell^p(G_{\hat{\mathcal{F}}, D^+}), \tag{3.2}$$

where $G_{\hat{\mathcal{F}}, D^+} = \{g \in G \mid \text{dist}\,(g\hat{\mathcal{F}}, D^+) \geq 1\}$. Let $\mathcal{O} := X \setminus D^+$ then $Au = 0$ on $\mathcal{O}$ and moreover, the set $G^{\mathcal{O}} = \{g \in G \mid g\hat{\mathcal{F}} \cap D^+ = \emptyset\}$ contains $G_{\hat{\mathcal{F}}, D^+}$. Due to the Schauder estimate of Proposition 5.5, for any $g \in G_{\hat{\mathcal{F}}, D^+}$, one has

$$\|u\|_{H^m(g\mathcal{F})} \lesssim \|u\|_{L^2(g\hat{\mathcal{F}})}. \tag{3.3}$$

By (3.2) and (3.3),

$$\{\|u\|_{H^m(g\mathcal{F})} \cdot \langle g \rangle^{-N}\}_{G_{\mathcal{F}, D^+}} \in \ell^p(G_{\mathcal{F}, D^+}). \tag{3.4}$$

Using Remark 2.2 again, this shows that $u \in L(\mu, A^p_{m,N})$, which in turn proves (3.1).

Definition 3.3 We denote by $(A^p_{m,N})^*$ the elliptic operator A^* with the domain

$$\text{Dom}\,(A^p_{m,N})^* = \{v \in V^{p'}_{m,-N}(X) \mid A^*v \in C^\infty_c(X)\},$$

where $1/p + 1/p' = 1$. In other words, $(A^p_{m,N})^* = (A^*)^{p'}_{m,-N}$.

We also define

$$\text{Dom}'\,A^p_{m,N} = \text{Dom}'\,(A^p_{m,N})^* := C^\infty_c(X).$$

Clearly, $C^\infty_c(X) \subseteq \text{Dom}\,A^p_{m,N}$ and $\text{Dom}\,(A^p_{m,N})^* \subseteq C^\infty(X)$ (see (1.19)).

This step shows that instead of dealing with A^p_N, it suffices to work with $A^p_{m,N}$ and its "adjoint" $(A^p_{m,N})^*$ (in the sense of Sect. 1.8.1), which are easier to deal with.

In the next steps, we will apply Theorem 1.26 to the operators $A^p_{m,N}$ and $(A^p_{m,N})^*$.

Step 2 In order to apply Theorem 1.26, we need to check properties $(\mathcal{P}1) - (\mathcal{P}4)$. The first three, $(\mathcal{P}1) - (\mathcal{P}3)$ hold by definition. To show that $(\mathcal{P}4)$ also holds, let us consider $u \in \mathrm{Dom}\, A_{m,N}^p$ and $v \in \mathrm{Dom}\,(A_{m,N}^p)^*$.

Note that since the operator A is G-periodic,

$$\|Au\|_{L^2(g\mathcal{F})} \lesssim \|u\|_{H^m(g\mathcal{F})} \tag{3.5}$$

and

$$\|A^*v\|_{L^2(g\mathcal{F})} \lesssim \|v\|_{H^m(g\mathcal{F})} \tag{3.6}$$

for any $g \in G$. Now, by Hölder's inequality, we have

$$\left| \sum_{g \in G} \langle Au, v \rangle_{L^2(g\mathcal{F})} \right| \leq \sum_{g \in G} \left| \langle Au, v \rangle_{L^2(g\mathcal{F})} \right| \leq \sum_{g \in G} \|Au\|_{L^2(g\mathcal{F})} \cdot \|v\|_{L^2(g\mathcal{F})}$$

$$\leq \|\{\|Au\|_{L^2(g\mathcal{F})}\langle g \rangle^{-N}\}_{g \in G}\|_{\ell^p(G)} \cdot \|\{\|v\|_{L^2(g\mathcal{F})}\langle g \rangle^{N}\}_{g \in G}\|_{\ell^{p'}(G)}$$

$$\lesssim \|\{\|u\|_{H^m(g\mathcal{F})}\langle g \rangle^{-N}\}_{g \in G}\|_{\ell^p(G)} \cdot \|\{\|v\|_{H^m(g\mathcal{F})}\langle g \rangle^{N}\}_{g \in G}\|_{\ell^{p'}(G)} \tag{3.7}$$

Similarly,

$$\left| \sum_{g \in G} \langle u, A^*v \rangle_{L^2(g\mathcal{F})} \right| \leq \|\{\|u\|_{H^m(g\mathcal{F})}\langle g \rangle^{-N}\}_{g \in G}\|_{\ell^p(G)} \cdot \|\{\|v\|_{H^m(g\mathcal{F})}\langle g \rangle^{N}\}_{g \in G}\|_{\ell^{p'}(G)}$$

Hence, both $\langle A_{m,N}^p u, v \rangle$ and $\langle u, (A_{m,N}^p)^* v \rangle$ are well-defined.

Our goal is to show that these two quantities are equal. To do this, for each $r \in \mathbb{N}$, we define

$$G_r := \{g \in G \mid |g| \geq r\},$$

and by $\mathcal{F}_r$ the union of all shifts of $\mathcal{F}$ by deck group elements whose word lengths do not exceed r, i.e.,

$$\mathcal{F}_r := \bigcup_{g \notin G_{r+1}} g\mathcal{F} = \bigcup_{|g| \leq r} g\mathcal{F}.$$

Obviously, $\mathcal{F}_r \Subset \mathcal{F}_{r+1}$ for any $r \in \mathbb{N}$ and the union of these subsets $\mathcal{F}_r$ is the whole covering X. Let $\phi_r \in C_c^\infty(X)$ be a cut-off function such that $\phi_n = 1$ on $\mathcal{F}_r$ and $\mathrm{supp}\,\phi_r \Subset \mathcal{F}_{r+1}$. Furthermore, all derivatives of ϕ_r are uniformly bounded with respect to r. In particular, the following estimates hold for any smooth function w on X and any $g \in G$:

$$\|\phi_r w\|_{H^m(g\mathcal{F})} \lesssim \|w\|_{H^m(g\mathcal{F})}, \quad \|(1 - \phi_r)w\|_{H^m(g\mathcal{F})} \lesssim \|w\|_{H^m(g\mathcal{F})}. \tag{3.8}$$

Let $u_r := \phi_r u$ and $v_r := \phi_r v$. Since u_r and v_r are compactly supported smooth functions on X, $\langle Au_r, v_r \rangle = \langle u_r, A^* v_r \rangle$. Therefore, it is enough to show that

$$\langle Au_r, v_r \rangle \to \langle Au, v \rangle \quad \text{and} \quad \langle u_r, A^* v_r \rangle \to \langle u, A^* v \rangle, \tag{3.9}$$

as $r \to \infty$. By symmetry, we only need to show the first part of (3.9). We use the triangle inequality to reduce (3.9) to checking that

$$\lim_{r \to \infty} \langle A(u - u_r), v \rangle = \lim_{r \to \infty} \langle Au_r, (v - v_r) \rangle = 0. \tag{3.10}$$

We repeat the argument of (3.7) for the pairs of functions $((1 - \phi_r)u, v)$ and $(\phi_r u, (1 - \phi_r)v)$, and then use (3.8) to derive

$$|\langle Au_r, (v - v_r) \rangle| + |\langle A(u - u_r), v \rangle|$$

$$\leq \sum_{g \in G} \left|\langle A(\phi_r u), (1 - \phi_r)v \rangle_{L^2(g\mathcal{F})}\right| + \left|\langle A((1 - \phi_r)u), v \rangle_{L^2(g\mathcal{F})}\right|$$

$$= \sum_{|g| \geq r+1} \left|\langle A(\phi_r u), (1 - \phi_r)v \rangle_{L^2(g\mathcal{F})}\right| + \left|\langle A((1 - \phi_r)u), v \rangle_{L^2(g\mathcal{F})}\right|$$

$$\lesssim \|\{\|\phi_r u\|_{H^m(g\mathcal{F})}\langle g \rangle^{-N}\}_{g \in G_{r+1}}\|_{\ell^p(G_{r+1})} \cdot \|\{\|(1 - \phi_r)v\|_{H^m(g\mathcal{F})}\langle g \rangle^N\}_{g \in G_{r+1}}\|_{\ell^{p'}(G_{r+1})}$$

$$+ \|\{\|(1 - \phi_r)u\|_{H^m(g\mathcal{F})}\langle g \rangle^{-N}\}_{g \in G_{r+1}}\|_{\ell^p(G_{r+1})} \cdot \|\{\|v\|_{H^m(g\mathcal{F})}\langle g \rangle^N\}_{g \in G_{r+1}}\|_{\ell^{p'}(G_{r+1})}$$

$$\lesssim \|\{\|u\|_{H^m(g\mathcal{F})}\langle g \rangle^{-N}\}_{g \in G_{r+1}}\|_{\ell^p(G_{r+1})} \cdot \|\{\|v\|_{H^m(g\mathcal{F})}\langle g \rangle^N\}_{g \in G_{r+1}}\|_{\ell^{p'}(G_{r+1})}. \tag{3.11}$$

Since $u \in V_{m,N}^p(X)$ and $v \in V_{m,-N}^{p'}(X)$, it follows that as $r \to \infty$, either

$$\|\{\|u\|_{H^m(g\mathcal{F})}\langle g \rangle^{-N}\}_{g \in G_{r+1}}\|_{\ell^p(G_{r+1})}$$

or

$$\|\{\|v\|_{H^m(g\mathcal{F})}\langle g \rangle^N\}_{g \in G_{r+1}}\|_{\ell^{p'}(G_{r+1})}$$

converges to zero (depending on either p or p' is finite), while the other one stays bounded. Thus, we have

$$\lim_{r \to \infty} \|\{\|u\|_{H^m(g\mathcal{F})}\langle g \rangle^{-N}\}_{g \in G_{r+1}}\|_{\ell^p(G_{r+1})} \cdot \|\{\|v\|_{H^m(g\mathcal{F})}\langle g \rangle^N\}_{g \in G_{r+1}}\|_{\ell^{p'}(G_{r+1})} = 0.$$

This fact and (3.11) imply (3.10). Hence, the property $(\mathcal{P}4)$ holds for $A^p_{m,N}$ and $(A^p_{m,N})^*$.

Step 3 Clearly,

$$\operatorname{Ker} A^p_{m,N} = \{u \in \operatorname{Dom} A^p_{m,N} \mid Au = 0\} = V^p_{m,N}(A) = V^p_N(A). \tag{3.12}$$

The latter equality is due to Schauder estimates (see (3.3) in Step 1). Also,

$$\operatorname{Ker}(A^p_{m,N})^* = V^{p'}_{m,-N}(A^*) = V^{p'}_{-N}(A^*) = 0 \tag{3.13}$$

according to Theorem 1.14. Hence, the kernels of $A^p_{m,N}$ and $(A^p_{m,N})^*$ are finite dimensional.

To prove that $A^p_{m,N}$ is Fredholm on X, we only need to show that

$$\operatorname{Im} A^p_{m,N} = C^\infty_c(X) = \left(\operatorname{Ker}(A^p_{m,N})^*\right)^\circ. \tag{3.14}$$

Given any $f \in C^\infty_c(X)$, we want to find a solution u of the equation $Au = f$ such that $u \in V^p_N(X)$. If such a solution u is found, then automatically u is in $V^p_{m,N}(X)$ by the same argument as in Step 1 and the fact that $Au = 0$ on the complement of the compact support of f. Thus, f must belong to the range of $A^p_{m,N}$ and the proof is then finished. So our remaining task is to find such a solution u. This can be done as follows: First, we pick a cut-off function η_r such that $\eta = 1$ around k_r and $\operatorname{supp} \eta_r \Subset V_r$, where V_r is the neighborhood of k_r appearing in *Assumption $\mathcal{A}$*. Define

$$\eta := \sum_{r=1}^{\ell} \eta_r \tag{3.15}$$

and notice that the Floquet transform $\mathbf{F}f$ is smooth in (k, x) since $f \in C^\infty_c(X)$. We decompose $\mathbf{F}f = \eta\mathbf{F}f + (1 - \eta)\mathbf{F}f$. Since the operator $A(k)$ is invertible when $k \notin F_{A,\mathbb{R}}$, the operator function

$$\widehat{u_0}(k) := A(k)^{-1}((1 - \eta(k))\mathbf{F}f(k)) \tag{3.16}$$

is well-defined and smooth in (k, x). By Theorem 5.4, the function $u_0 := \mathbf{F}^{-1}\widehat{u_0}$ has rapid decay. We recall that when $k \in V_r$, the Riesz projection $\Pi_r(k)$ is defined in *Assumption $\mathcal{A}$*. Clearly,

$$0 \notin \sigma(A(k)_{|R(1-\Pi_r(k))}), \tag{3.17}$$

where we use the notation $R(T)$ for the range of an operator T. Now the operator function

$$\widehat{v_r}(k) := \eta_r(k)(A(k)_{|R(1-\Pi_r(k))})^{-1}(1 - \Pi_r(k))\mathbf{F}f(k) \tag{3.18}$$

is also smooth and thus the function $v_r := \mathbf{F}^{-1}\widehat{v_r}$ has rapid decay by Theorem 5.4. In particular, $u_0, v_r (1 \leq r \leq \ell)$ are in the space $V_0^\infty(X)$.

Let us fix $1 \leq r \leq \ell$. For any $k \in V_r \setminus \{k_r\}$, due to $(\mathcal{A}4)$, we can define the operator function

$$\widehat{w_r}(k) := \eta_r(k)(A(k)_{|R(\Pi_r(k))})^{-1}\Pi_r(k)\mathbf{F}f(k), \tag{3.19}$$

which is in the range of $\Pi_r(k)$. By expanding $\Pi_r(k)\mathbf{F}f(k)$ in terms of the basis $(f_j(k))_{1 \leq j \leq m_r}$, one sees that

$$\|\widehat{w_r}(k)\|_{L_k^2(X)} \lesssim \max_{1 \leq j \leq m_r} \|A_r(k)^{-1}f_j(k)\|_{L_k^2(X)} \cdot \|\mathbf{F}f(k)\|_{L_k^2(X)}.$$

From this and the integrability condition in $(\mathcal{A}2)$, we obtain

$$\int_{\mathbb{T}^d} \|\widehat{w_r}(k)\|_{L_k^2(X)}\mathrm{d}k \lesssim \int_{V_r \setminus \{k_r\}} \|\mathbf{F}f(k)\|_{L_k^2(X)} \cdot \|(A_r(k)^{-1}f_j(k))_{1 \leq j \leq m_r}\|_{\ell^\infty(\mathbb{C}^{m_r})}\mathrm{d}k$$

$$\lesssim \sup_{k \in V_r} \|\mathbf{F}f(k)\|_{L_k^2(X)} \cdot \int_{V_r \setminus \{k_r\}} \|A_r(k)^{-1}\|_{\mathcal{L}(\mathbb{C}^{m_r})}\mathrm{d}k < \infty.$$

Hence, $\widehat{w_r} \in L^1(\mathbb{T}^d, \mathcal{E}^0)$.

Summing up, the function $\widehat{u} := \widehat{u}_0 + \sum_{1 \leq r \leq \ell}(\widehat{v_r} + \widehat{w_r})$ belongs to $L^1(\mathbb{T}^d, \mathcal{E}^0)$, and moreover, it satisfies the equation

$$A(k)\widehat{u}(k) = A(k)\widehat{u}_0(k) + \sum_{1 \leq r \leq \ell} A(k)(\widehat{v_r}(k) + \widehat{w_r}(k))$$

$$= (1 - \eta(k))\mathbf{F}f(k) + \sum_{1 \leq r \leq \ell} \eta_r(k)\mathbf{F}f(k) = \mathbf{F}f(k). \tag{3.20}$$

From the above equality, $\widehat{u}(k, x)$ is smooth in x for each quasimomentum k. We define $u := \mathbf{F}^{-1}\widehat{u}$ by using the formula (5.13). According to Lemma 5.2, $u \in L_{loc}^2(X)$. For any $\phi \in C_c^\infty(X)$, we can use Fubini's theorem to get

$$\langle u, A^*\phi \rangle_{L^2(X)} = \int_X \mathbf{F}^{-1}\widehat{u}(k, x) \cdot A^*\phi(x)\mathrm{d}\mu_X(x)$$

$$= \frac{1}{(2\pi)^d} \int_{\mathbb{T}^d} \int_X \widehat{u}(k, x) \cdot A^*\phi(x)\mathrm{d}\mu_X(x)\mathrm{d}k$$

$$= \frac{1}{(2\pi)^d} \int_{\mathbb{T}^d} \int_X A\widehat{u}(k, x) \cdot \phi(x) \mathrm{d}\mu_X(x) \mathrm{d}k$$

$$= \frac{1}{(2\pi)^d} \int_{\mathbb{T}^d} \int_X A(k)\widehat{u}(k, x) \cdot \phi(x) \mathrm{d}\mu_X(x) \mathrm{d}k$$

$$= \frac{1}{(2\pi)^d} \int_{\mathbb{T}^d} \int_X \mathbf{F}f(k, x) \cdot \phi(x) \mathrm{d}\mu_X(x) \mathrm{d}k$$

$$= \langle f, \phi \rangle_{L^2(X)}.$$

Hence, u is a weak solution of the inhomogeneous equation $Au = f$ on X. Elliptic regularity then implies that u is a classical solution and therefore, $u \in V_0^\infty(X)$ due to Lemma 5.2 again. If either $N \geq 0$ when $p = \infty$ or $N > d/p$ when $p \in [1, \infty)$, we always have $V_0^\infty(X) \subseteq V_N^p(X)$. Thus, this shows that $A_{m,N}^p$ is a Fredholm operator on X.

Step 4   Due to considerations in Step 2 and Step 3, the operator $A_{m,N}^p$ satisfies the assumption of Theorem 1.26. Then the Liouville-Riemann-Roch inequality (2.2) follows immediately from (3.1) and Theorem 1.26.

Remark 3.1   Assumption $(\mathcal{A}2)$ is needed to guarantee the validity of the Liouville-Riemann-Roch inequality (2.2) (at least when $p = \infty$ and $N = 0$). Indeed, consider $-\Delta$ in $\mathbb{R}^2$. Let μ be the point divisor $(\{0\}, L, \emptyset, 0)$, where $L = \mathbb{C}\delta_0$. It is not difficult to see that the space $L_\infty(\mu, -\Delta, 0)$ contains only constant functions, since the standard fundamental solution $u_0(x) = -\frac{1}{2\pi} \ln|x|$ is not bounded at infinity. Clearly, $L_1(\mu^{-1}, -\Delta, 0)$ is trivial. Hence, we have:

$$\dim L_\infty(\mu, -\Delta, 0) = 1 < 2 = \deg_{-\Delta}(\mu) + \dim V_0^\infty(-\Delta) + \dim L_1(\mu^{-1}, -\Delta, 0). \tag{3.21}$$

3.3   Proof of Theorem 2.5

According to the Step 3 of the proof of Theorem 2.2, the operator $A_{m,N}^p$ is Fredholm on X and

$$\operatorname{Im} A_{m,N}^p = C_c^\infty(X) = \operatorname{Dom}'(A_{m,N}^p)^*. \tag{3.22}$$

Now we can apply Corollary 1.27 to finish the proof of the equality (2.3). The upper bound estimate (2.4) follows from (2.3) and the trivial inclusion $L_p(\mu, A, N) \subseteq L_p(\mu^+, A, N)$.

3.3.1 Proof of Proposition 2.7

a. It suffices to prove the statement for the case $p = \infty$. If $r \geq 0$, we define a point divisor $\mu_r := (\emptyset, 0; \{0\}, L_r^-)$, where

$$L_r^- := \left\{ \sum_{|\alpha| \leq r} c_\alpha \partial^\alpha \delta(\cdot - 0) \mid c_\alpha \in \mathbb{C} \right\}. \tag{3.23}$$

Let us consider the function $v_\alpha(x) := \partial^\alpha(|x|^{2-d})$ for each multi-index α such that $|\alpha| > N + 2$. It is clear that $|v_\alpha(x)| \lesssim |x|^{-|\alpha|-d+2}$ for $x \neq 0$. Therefore,

$$\sum_{g \in \mathbb{Z}^d} \|v_\alpha\|_{L^2([0,1)^d + g)} \cdot \langle g \rangle^N \lesssim \sum_{g \in \mathbb{Z}^d} \langle g \rangle^{-|\alpha|-d+2+N} < \infty. \tag{3.24}$$

Since $|x|^{2-d}$ is a fundamental solution of $-\Delta$ on $\mathbb{R}^d$ (up to some multiplicative constant), v_α belongs to the space $L_1(\mu_r^{-1}, -\Delta_{|\mathbb{R}^d}, -N)$ provided that $N + 2 < |\alpha| \leq r$. Now let us pick multi-indices $\alpha_1, \ldots, \alpha_{r-N-2}$ such that $|\alpha_j| = N+2+j$ for any $1 \leq j \leq r - N - 2$. By homogeneity, the functions v_{α_j} are linearly independent as smooth functions on $\mathbb{R}^d \setminus \{0\}$. By letting $r \to \infty$, this proves the statement a.

b. We define $\mu_0 := (\emptyset, 0; D^-, L^-)$. Now suppose the contrary, that for any $M > 0$, the space $L_{p'}(\mu_M^{-1}, A^*, -N)$ is non-trivial for some rigged divisor $\mu_M = (D^+, L_M^+; D^-, L^-)$ such that $\mathcal{L}_M^+ \subseteq L_M^+$. Note that $L_{p'}(\mu_M^{-1}, A^*, -N)$ is a subspace of $L_{p'}(\mu_0^{-1}, A^*, -N)$. It follows from Proposition 2.5 that $L_{p'}(\mu_0^{-1}, A^*, -N)$ is a finite dimensional vector space and thus, we equip it with any norm $\| \cdot \|$. Thus, there is a sequence $\{u_M\}_{M \in \mathbb{N}}$ in $L_{p'}(\mu_0^{-1}, A^*, -N)$ such that $\|u_M\| = 1$ and $(u_M, L_M^+) = 0$. In particular, $(u_M, \mathcal{L}_M^+) = 0$ and therefore, $\partial^\alpha u_M(x_0) = 0$ for any $0 \leq |\alpha| \leq M$. By passing to a subsequence if necessary, there exists $v \in L_{p'}(\mu_0^{-1}, A^*, -N)$ for which $\lim_{M \to \infty} \|u_M - v\| = 0$. It is clear that

$$\|w\|_{C^M(K)} \lesssim \|w\| \text{ for any } w \text{ in } L_{p'}(\mu_0^{-1}, A^*, -N), M \geq 0, \text{ and compact subset}$$

$K \Subset X \setminus D^-$. Hence, for any multi-index α, $\partial^\alpha v(x_0) = \lim_{M \to \infty} \partial^\alpha u_M(x_0) = 0$.

As a local smooth solution of A^*, v must vanish on $X \setminus D^-$ due to the strong unique continuation property of A^*. Consequently, $v = 0$ as an element in $L_{p'}(\mu_0^{-1}, A^*, -N)$ and this gives us a contradiction with $\|v\| = 1$. This completes our proof.

Remark 3.2

(i) There are large classes of elliptic operators with the strong unique continuation property, e.g. elliptic operators of second-order with smooth coefficients and elliptic operators with real analytic coefficients.

(ii) Note that the finiteness of the real Fermi surface $F_{A,\mathbb{R}}$ would imply the weak unique continuation property of A^*, i.e., A^* does not have any non-zero compactly supported solution (see e.g., [41]). We do not know whether the first statement of Proposition 2.7 still holds without the strong unique continuation property requirement for A^*.

3.4 Proof of Theorem 2.8

The proof is similar to the one of Theorem 2.2, except for Step 3, where it needs a minor modification. We keep the same considerations and notions as in Step 3.

The goal here is to prove the solvability in $V_N^p(X)$ of the equation $Au = f$, where $f \in C_c^\infty(X)$. Under the assumption (2.5), the functions $\widehat{w}_r$ ($1 \leq r \leq \ell$) defined in Step 3 belong to $L^2(\mathbb{T}^d, \mathcal{E}^0)$. Thus, $\widehat{u} \in L^2(\mathbb{T}^d, \mathcal{E}^0)$ and then by Theorem 5.4, u is in $L^2(X)$ and $Au = f$. This means that $u \in V_0^2(X)$.

If $p \geq 2, N \geq 0$, the inclusion $V_0^2(X) \subseteq V_N^2(X)$ is obvious, while if $p \in [1, 2), N > (1/p - 1/2)d$, one can use Hölder's inequality to obtain the inclusion $V_0^2(X) \subset V_N^p(X)$.

This completes the proof of the first statement. In particular, when $p = 2, N = 0$, both operators A_0^2 and $(A_0^2)^* = (A^*)_0^2$ are Fredholm. Therefore, we obtain the equality (2.7), since $\dim V_0^2(A) = \dim V_0^2(A^*) = 0$ according to Theorem 1.14 (a).

Remark 3.3

(a) The integrability of $\|A_r(k)^{-1}\|_{\mathcal{L}(\mathbb{C}^{mr})}^2$ is important for validity of Theorem 2.8. Indeed, let us consider $A = -\Delta$ on $\mathbb{R}^d$ ($d < 5$) and the point divisor μ representing a simple pole at 0. Then the fundamental solution $c_d|x|^{2-d}$ does not belong to $L_2(\mu, -\Delta, 0)$ and thus, $\dim L_2(\mu, -\Delta, 0) = 0 < 1 = \deg_{-\Delta}(\mu)$. Therefore, the equalities (2.6) and (2.7) do not hold in this case.

(b) Under the assumption of Theorem 2.8, the Liouville-Riemann-Roch inequality (2.2) holds for **any** $N \geq 0$ if and only if $p \geq 2$. Indeed, suppose that $d \geq 5$ and $p < 2$, then $(2 - d)p \geq -d$ and therefore,

$$\int_{|x| \geq 1} |x|^{(2-d)p} dx = \infty. \tag{3.25}$$

This implies that $L_p(\mu, -\Delta, 0) = \{0\}$, where μ is the same point divisor mentioned in the previous remark. So (2.2) fails, since,

$$\dim L_p(\mu, -\Delta, 0) = 0 < 1 = \deg_{-\Delta}(\mu). \tag{3.26}$$

3.4.1 Proof of Proposition 2.9

We evoke the operators $A_{m,0}^2$ and $(A_{m,0}^2)^*$ and their corresponding domains from the proof of Theorem 2.2. Now we recall from our discussion in Sect. 1.8 the notations of the operators

$$\widetilde{A_{m,0}^2} : \Gamma(X, \mu^{-1}, A_{m,0}^2) \to \tilde{\Gamma}_{\mu^{-1}}(X, A_{m,0}^2) \tag{3.27}$$

and

$$\widetilde{(A_{m,0}^2)^*} : \Gamma(X, \mu, (A_{m,0}^2)^*) \to \tilde{\Gamma}_\mu(X, (A_{m,0}^2)^*), \tag{3.28}$$

which are extensions of $A_{m,0}^2$ and $(A_{m,0}^2)^*$ with respect to the divisors μ^{-1} and μ, correspondingly. From (2.7) and Remark 1.10, we obtain the duality

$$L_2(\mu, A^*, 0)^\circ = (\mathrm{Ker}\,\widetilde{(A_{m,0}^2)^*})^\circ = \mathrm{Im}\,\widetilde{A_{m,0}^2}. \tag{3.29}$$

To put it differently, f is orthogonal to $\tilde{L}$ and $L_2(\mu, A^*, 0)$ if and only if $f = Au$ for some u in the space

$$\Gamma(X, \mu^{-1}, A_{m,0}^2) = \{v \in H^m(X) \mid Av \in C_c^\infty(X), \langle v, L \rangle = 0\}. \tag{3.30}$$

This proves the equivalence of (i) and (ii).

3.4.2 Proof of Proposition 2.10

For $\ell \in \mathbb{N}$, let us choose ℓ distinct points $z_1, \ldots, z_\ell$ in $\mathbb{R}^d$ and define $\mu_\ell := (\emptyset, 0; D^-, L^-)$, where $D^- = \{z_1, \ldots, z_\ell\}$ and

$$L^- = \left\{ \sum_{1 \le j \le \ell} \sum_{1 \le \alpha \le d} c_{j\alpha} \frac{\partial}{\partial x_\alpha} \delta(x - z_j) \mid c_{j\alpha} \in \mathbb{C} \right\}. \tag{3.31}$$

In terms of the notations in Example 1.1, $k = 0, l = \ell$.

Let us recall now the spaces $L(\mu, -\Delta)$ and $L(\mu^{-1}, -\Delta)$ from Example 1.1. By definition, $L_\infty(\mu, -\Delta, 0) = \mathbb{C}$. Hence,

$$\dim L_\infty(\mu, -\Delta, 0) = 1 = \dim L(\mu, -\Delta) + \dim V_0^\infty(-\Delta). \tag{3.32}$$

On the other hand, if $v \in L_1(\mu^{-1}, -\Delta, 0)$, then

$$\lim_{R \to \infty} \sum_{|g| \geq R} \|v\|_{L^2([0,1)^d+g)} = 0. \tag{3.33}$$

Hence, $\|v\|_{L^2([0,1)^d+g)} \to 0$ as $|g| \to \infty$. Using elliptic regularity, this is equivalent to $\lim_{|x| \to \infty} v(x) = 0$. Thus, $L_1(\mu^{-1}, -\Delta, 0)$ is a subspace of $L(\mu^{-1}, -\Delta)$. Define

$$v_{j\alpha}(x) := \frac{\partial}{\partial x_\alpha} |x - z_j|^{2-d}. \tag{3.34}$$

Then $v_{j\alpha} \in L(\mu^{-1}, -\Delta)$ (see Example 1.1).[1] We now claim that $v_{j\alpha} \notin L_1(\mu^{-1}, -\Delta, 0)$ for any $1 \leq j \leq \ell$ and $1 \leq \alpha \leq d$. Suppose this is not true:

$$v_{j\alpha}(x) = (2 - d)\frac{(x_\alpha - (z_j)_\alpha)}{|x - z_j|^d} \in L_1(\mu^{-1}, -\Delta, 0). \tag{3.35}$$

This implies that for some $R > 0$, we have

$$V_{\alpha,R} := \sum_{g \in \mathbb{Z}^d, |g| \geq R} \left(\int_{[0,1)^d+g} \frac{|x_\alpha - (z_j)_\alpha|^2}{|x - z_j|^{2d}} dx \right)^{1/2} < \infty.$$

But this leads to a contradiction, since

$$V_{\alpha,R} \gtrsim \sum_{g \in \mathbb{Z}^d, |g| \geq R} \frac{|g_\alpha|}{|g|^d} \geq \sum_{g \in \mathbb{Z}^d, g_\alpha \neq 0, |g| \geq R} \frac{1}{|g|^d} = \infty. \tag{3.36}$$

From this and linear independence of functions $v_{j\alpha}$ as smooth functions on $\mathbb{R}^d \setminus D^-$, it follows that

$$\dim L_1(\mu^{-1}, -\Delta, 0) \leq \dim L(\mu^{-1}, -\Delta) - d\ell. \tag{3.37}$$

From (1.37), (3.32) and (3.37), we get

$$d\ell + \dim L_1(\mu^{-1}, -\Delta, 0) + \deg_{-\Delta}(\mu) + \dim V_0^\infty(-\Delta)$$
$$\leq \dim L(\mu^{-1}, -\Delta) + \deg_{-\Delta}(\mu) + \dim V_0^\infty(-\Delta) = \dim L_\infty(\mu, -\Delta, 0), \tag{3.38}$$

which yields the inequality (2.8).

---

[1] In physics, the functions $v_{j\alpha}$ in $L(\mu^{-1}, -\Delta)$ are potentials of dipoles located at the equilibrium positions z_j.

Remark 3.4

(i) Note that the examples from Proposition 2.10 also show that the Liouville-Riemann-Roch inequality can be strict in some other cases as well.

Case 1: $p = \infty$ and $N \geq 0$

If $(d - 1)\ell + 1 \geq \dim V_N^\infty(-\Delta)$, one obtains from (3.38) that

$$\dim L_1(\mu^{-1}, -\Delta, -N) + \deg_{-\Delta}(\mu) + \dim V_N^\infty(-\Delta) + \ell$$
$$\leq \dim L_1(\mu^{-1}, -\Delta, 0) + \deg_{-\Delta}(\mu) + \dim V_0^\infty(-\Delta) + d\ell \qquad (3.39)$$
$$\leq \dim L_\infty(\mu, -\Delta, N).$$

Case 2: $1 \leq p < \infty$ and $N > d/p$

Note that $\dim L_p(\mu, -\Delta, N) \geq 1$, since this space contains constant solutions. Each function $v_{j\alpha}$ does not belong to the space $L_{p'}(\mu^{-1}, -\Delta, -N)$. In fact, for $R > 2|z_j|$ large enough and $p > 1$, we get

$$\sum_{g \in \mathbb{Z}^d, |g| > R} \|v_{j\alpha}\|_{L^2([0,1)^d + g)}^{p'} \cdot \langle g \rangle^{p'N} \gtrsim \sum_{\min_{1 \leq l \leq d} g_l > R} \langle g \rangle^{p'(N-d)} = \infty. \qquad (3.40)$$

The case when $p = 1$ and $N > d - 1$ can be treated similarly.

Now, as in the proof of (3.38), we obtain the inequality

$$\dim L_{p'}(\mu^{-1}, -\Delta, -N) + \deg_{-\Delta}(\mu) + \dim V_N^p(-\Delta) + \ell$$
$$\leq \dim L(\mu^{-1}, -\Delta) - d\ell + \deg_{-\Delta}(\mu) + \dim V_N^p(-\Delta) + \ell \qquad (3.41)$$
$$\leq \dim L_p(\mu, -\Delta, N),$$

provided that $(d - 1)\ell + 1 \geq \dim V_N^p(-\Delta)$.

(ii) One can also modify our example in Proposition 2.10 to obtain examples of (2.8) in the case of point divisors. For instance, we could take the point divisor $\mu = (\emptyset, 0; D^-, L^-)$, where $D^- = \{z_1, \ldots, z_\ell\}$ and

$$L^- = \text{span}_{\mathbb{C}} \{\partial^\alpha \delta(x - z_j)\}_{1 \leq j \leq \ell, 0 \leq |\alpha| \leq 1}. \qquad (3.42)$$

Similarly,

$$L_\infty(\mu, -\Delta, 0) = L_1(\mu^{-1}, -\Delta, 0) = \{0\}. \qquad (3.43)$$

Moreover, $\deg_{-\Delta}(\mu) = -\ell(d + 1)$. Hence,

$$\dim L_\infty(\mu, -\Delta, 0) = (\ell(d + 1) - 1) + \deg_{-\Delta}(\mu) + \dim V_0^\infty(-\Delta)$$
$$+ \dim L_1(\mu^{-1}, -\Delta, 0). \qquad (3.44)$$

The method can also be easily adapted to providing examples of the inequality (2.8) when both the positive parts μ^+ and negative parts μ^- of the rigged divisors μ are non-trivial.

3.4.3 Proof of Proposition 2.11

We can assume w.l.o.g that $x_0 = 0$. Let us now fix a pair (p, N) as in the assumption of the statement. We recall the notations of the operator $A_{m,N}^p$ and its corresponding domain Dom $A_{m,N}^p$ from Definition 3.2.

In order to prove the statement of the proposition, we will apply Corollary 1.27 to the operator $P := A_{m,N}^p$.

From our assumption on the operator A and the pair (p, N) and from the conclusion of Step 3 of the proof of Theorem 2.2, we only need to show the following claim: If u is a smooth function on $\mathbb{R}^d$ such that $|u(x)| \lesssim \langle x \rangle^N$ and $\langle Au, \tilde{L}^- \rangle = 0$, then there is a polynomial v of degree M_0 satisfying $Av = 0$ and $\langle u - v, L^- \rangle = 0$. Indeed, if this claim holds true, v will belong to the space Dom $A_{m,N}^p$ due to our condition on p and N. This will fulfill all the necessary assumptions of Corollary 1.27 in order to apply it.

To prove the claim, we first introduce the following polynomial:

$$v(x) := \sum_{M_1 \leq |\alpha| \leq M_0} \frac{\partial^\alpha u(0)}{\alpha!} x^\alpha. \tag{3.45}$$

Hence, $\langle v - u, g \rangle = 0$ if $g = \partial^\alpha \delta(\cdot - 0)$ and $M_1 \leq |\alpha| \leq M_0$. Let $a(\xi)$ be the symbol of the constant-coefficient differential operator $A(x, D)$, i.e.,

$$A = A(x, D) = \sum_{|\alpha|=m} \frac{1}{\alpha!} \partial_\xi^\alpha a(0) D^\alpha. \tag{3.46}$$

Define $\tilde{M}_j := \max\{0, M_j - m\}$ for $j \in \{0, 1\}$. A straightforward calculation gives:

$$A(x, D)v(x) = i^m \sum_{|\alpha|=m} \sum_{|\beta|=\tilde{M}_1}^{\tilde{M}_0} \frac{1}{\alpha!} \frac{1}{\beta!} \partial_\xi^\alpha a(0) \partial_x^{\alpha+\beta} u(0) \cdot x^\beta$$

$$= \sum_{|\beta|=\tilde{M}_1}^{\tilde{M}_0} \frac{1}{\beta!} \left(\sum_{|\alpha|=m} \frac{1}{\alpha!} \partial_\xi^\alpha a(0) \cdot D^\alpha (\partial_x^\beta u)(0) \right) \cdot x^\beta \tag{3.47}$$

$$= \sum_{|\beta|=\tilde{M}_1}^{\tilde{M}_0} \frac{1}{\beta!} A \partial^\beta u(0) \cdot x^\beta.$$

Because $\partial^\beta \delta(\cdot - 0) \in \tilde{L}^-$ when $\tilde{M}_1 \leq |\beta| \leq \tilde{M}_0$, we obtain $A \partial^\beta u(0) = \partial^\beta A u(0) = 0$ for such multi-indices β. Now we conclude that $Av = 0$, which proves our claim.

Remark 3.5

(i) If $d > m$ in Proposition 2.11, then any elliptic real-constant-coefficient homogeneous differential operator A of order m on $\mathbb{R}^d$ satisfies Assumption $\mathcal{A}$. Notice that m must be even. Since $F_{A,\mathbb{R}} = \{0\}$, it is not hard to see from Theorem 1.17 that

$$\dim V_N^p(A) = \dim V_N^p((-\Delta)^{m/2}). \tag{3.48}$$

In particular, if μ is the point divisor $x_0^{-(M_0+1)}$, the Liouville-Riemann-Roch formula becomes

$$\dim L_p(\mu, A, N) = \begin{cases} h_{d,[N]}^{(m)} - h_{d,M_0}^{(m)} & \text{if } p = \infty. \\ h_{d,\lfloor N-d/p \rfloor}^{(m)} - h_{d,M_0}^{(m)} & \text{otherwise,} \end{cases} \tag{3.49}$$

though this also has an elementary proof.

Here for $A, B, C \in \mathbb{N}$, we denote by $h_{A,B}^{(C)}$ the quantity $\binom{A+B}{A} - \binom{A+B-C}{A}$, where we adopt the agreement in Definition 1.16.

(ii) As a special case of Proposition 2.11, the Liouville-Riemann-Roch equality for the Laplacian operator could occur when μ^- is non-trivial (compare with Theorem 2.5). As we have seen, the corresponding spaces $L_{p'}(\mu^{-1}, -\Delta, -N)$ in this situation are trivial.

It is worth mentioning that it is possible to obtain the Liouville-Riemann-Roch equality in certain cases when the dimensions of the spaces $L_{p'}(\mu^{-1}, -\Delta, -N)$ can be arbitrarily large. For instance, let $p = \infty$ and $r \geq N + 3$, we define

$$\mu := (\emptyset, 0; D^-, L^-) \text{ with } D^- = \{0\} \tag{3.50}$$

and

$$L^- = \text{span}_{\mathbb{C}}\{\partial^\alpha \delta(\cdot - 0)\}_{|\alpha|=r}. \tag{3.51}$$

Then clearly $L_\infty(\mu, -\Delta, N) = V_N^\infty(-\Delta)$. From the proof of the second part of Proposition 2.7,

$$L_1(\mu^{-1}, -\Delta, -N) = \text{span}_{\mathbb{C}}\{\partial^\alpha(|x|^{2-d})\}_{|\alpha|=r}$$

$$= \{u \in C^\infty(\mathbb{R}^d \setminus \{0\}) \mid -\Delta u \in L^-, \lim_{|x| \to \infty} |u(x)| = 0\}. \tag{3.52}$$

By Theorem 1.26, it is easy to see that the dimension of this space is equal to the degree of the divisor μ^{-1} (see also Example 1.1). Thus,

$$\dim L_\infty(\mu, -\Delta, N) = \dim L_1(\mu^{-1}, -\Delta, -N) + \deg_{-\Delta}(\mu) + \dim V_N^\infty(-\Delta) \tag{3.53}$$

and as $r \to \infty$,

$$\dim L_1(\mu^{-1}, -\Delta, -N) = \binom{d+r-1}{d-1} - \binom{d+r-3}{d-1} \to \infty. \tag{3.54}$$

3.4.4   Proof of Corollary 2.12

In a similar manner to the proof of Corollary 2.9, for the rigged divisor μ_z the corresponding extension operator

$$\widetilde{(A_z)^2_{m,0}} : \Gamma(X, \mu_z, (A_z)^2_{m,0}) \to \tilde{\Gamma}_{\mu_z}(X, (A_z)^2_{m,0}) \tag{3.55}$$

is Fredholm. As in the proof of [72, Theorem 2], we can deduce the upper-semicontinuity of $\dim \operatorname{Ker} \widetilde{(A_z)^2_{m,0}}$ by using [72, Theorem 1] and [72, Theorem 3]. Since $\operatorname{Ker} \widetilde{(A_z)^2_{m,0}} = L_2(\mu_z, A_z, 0)$, this finishes our proof. The upper-semicontinuity of $z \mapsto \dim L_2(\mu_z^{-1}, A_z^*, 0)$ is proved similarly.

3.5   Proof of Theorem 2.16

The key of the proof is the following statement:

Lemma 3.1 *Let us consider $p_1, p_2 \in [1, \infty]$ and two positive functions φ_1 and φ_2 in $\mathcal{S}(G)$ such that we assume either one of the following two conditions:*

- *$p_1^{-1} + p_2^{-1} \geq 1$ and $\varphi_1\varphi_2$ is bounded on G.*
- *$p_1^{-1} + p_2^{-1} \leq 1$ and $\varphi_1^{-1}\varphi_2^{-1}$ is bounded on G.*

Then the Riemann-Roch formula holds:

$$\dim L_{p_1}(\mu, P, \varphi_1) = \deg_P(\mu) + \dim L_{p_2}(\mu^{-1}, P^*, \varphi_2), \tag{3.56}$$

where μ is any rigged divisor on $\mathcal{X}$.

Proof of Lemma 3.1 We follow the strategy of the proof of Theorem 2.2. As in Definition 3.2, for each $s \in \mathbb{R}$, $\varphi \in \mathcal{S}(G)$ and $p \in [1, \infty]$, let us introduce the following space

$$\mathcal{V}^p_{s,\varphi}(\mathcal{X}) := \{u \in C^\infty(\mathcal{X}) \mid \{\|u\|_{H^s(g\mathcal{F})} \cdot \varphi(g)\}_{g \in G} \in \ell^p(G)\}. \tag{3.57}$$

and we denote by $P^p_{m,\varphi}$ the operator P with the domain

$$\mathrm{Dom}\, P^p_{m,\varphi} := \{u \in \mathcal{V}^p_{m,\varphi}(\mathcal{X}) \mid Pu \in C^\infty_c(\mathcal{X})\}. \tag{3.58}$$

For the elliptic differential operator P^*, we also use the corresponding notations $(P^*)^p_{m,\varphi}$ and $\mathrm{Dom}(P^*)^p_{m,\varphi}$.

Now let us fix a pair of two real numbers (p_1, p_2) and a pair of two functions (φ_1, φ_2) satisfying the conditions of Lemma 3.1. From now on, we will consider the operator $P^{p_1}_{m,\varphi_1}$ and its "adjoint" $(P^{p_1}_{m,\varphi_1})^* := (P^*)^{p_2}_{m,\varphi_2}$. As before, we define $\mathrm{Dom}'\, P^{p_1}_{m,\varphi_1} = \mathrm{Dom}'\, (P^*)^{p_2}_{m,\varphi_2} = C^\infty_c(\mathcal{X})$.

Our goal is to verify the assumptions of Theorem 1.26 for the operator $P^{p_1}_{m,\varphi_1}$ and its adjoint $(P^*)^{p_2}_{m,\varphi_2}$. The proof also goes through four steps as in Theorem 2.2. We consider two cases.

Case 1 $p_1^{-1} + p_2^{-1} \geq 1$, $\varphi_1 \varphi_2 \lesssim 1$.
The proof of Step 1 stays exactly the same as before (see Remark 5.2).

For Step 2, the first three properties $(\mathcal{P}1) - (\mathcal{P}3)$ are immediate. For the property $(\mathcal{P}4)$, we want to show that whenever $u \in \mathrm{Dom}\, P^{p_1}_{m,\varphi_1}$ and $v \in \mathrm{Dom}(P^*)^{p_2}_{m,\varphi_2}$, then

$$\langle Pu, v \rangle = \langle u, P^*v \rangle. \tag{3.59}$$

Because P and P^* are C^∞-bounded, we can repeat the approximation procedure and obtain similar estimates from the proof of Theorem 2.2 for showing the identity (3.59) whenever $u \in \mathrm{Dom}\, P^{p_1}_{m,\varphi_1}$ and $v \in \mathrm{Dom}(P^*)^{p'_1}_{m,\varphi_1^{-1}}$. On the other hand, $\mathcal{V}^{p_2}_{m,\varphi_2}(\mathcal{X}) \subseteq \mathcal{V}^{p'_1}_{m,\varphi_1^{-1}}(\mathcal{X})$ and hence, $\mathrm{Dom}(P^*)^{p_2}_{m,\varphi_2} \subseteq \mathrm{Dom}(P^*)^{p'_1}_{m,\varphi_1^{-1}}$. With this inclusion, it is enough to conclude the property $(\mathcal{P}4)$ in this case, which finishes Step 2.

For Step 3, first, it is clear that the kernels of $P^{p_1}_{m,\varphi_1}$ and $(P^*)^{p_2}_{m,\varphi_2}$ are both trivial since P and P^* satisfy (SSP) (see Theorem 2.14). So the rest is to verify the Fredholm property of both operators $P^{p_1}_{m,\varphi_1}$ and $(P^*)^{p_2}_{m,\varphi_2}$, i.e., to prove that

$$\mathrm{Im}\, P^{p_1}_{m,\varphi_1} = \mathrm{Im}(P^*)^{p_2}_{m,\varphi_2} = C^\infty_c(\mathcal{X}). \tag{3.60}$$

Let us prove (3.60) for the operator $P^{p_1}_{m,\varphi_1}$ since the other identity is proved similarly. We denote by $G_P(x, y)$ the Green's function of P at the level $\lambda = 0$, i.e., $G_P(x, y)$ is the Schwartz kernel of the resolvent operator P^{-1}. It is known that $G_P(x, y) \in C^\infty(\mathcal{X} \times \mathcal{X} \setminus \Delta)$, where $\Delta = \{(x, x) \mid x \in \mathcal{X}\}$. Moreover, all of its derivatives have exponential decay off the diagonal (see [67, Theorem 2.2]). However, it is more convenient for us to use its L^2-norm version, i.e., [67, Theorem 2.3]: there exists $\varepsilon > 0$ such that for every $\delta > 0$ and every multi-indices α, β, one can find a constant $C_{\alpha\beta\delta} > 0$ such that

$$\int\limits_{x:d_{\mathcal{X}}(x,y)\geq\delta} |\partial_x^\alpha \partial_y^\beta G_P(x,y)|^2 \exp(\varepsilon d_{\mathcal{X}}(x,y)) \mathrm{d}\mu_{\mathcal{X}}(x) \leq C_{\alpha\beta\delta}. \tag{3.61}$$

Here the derivatives ∂_x^α, ∂_y^β are taken with respect to canonical coordinates and the constants $C_{\alpha\beta\delta}$ do not depend on the choice of such canonical coordinates. Note that these estimates (3.61) still work in the case of exponential growth of the volume of the balls on $\mathcal{X}$. Let $f \in C_c^\infty(\mathcal{X})$ and K be its compact support in $\mathcal{X}$. We introduce

$$u(x) := P^{-1} f(x) = \int_{\mathcal{X}} G_P(x,y) f(y) \mathrm{d}\mu_{\mathcal{X}}(y), \tag{3.62}$$

where $\mu_{\mathcal{X}}$ is the Riemannian measure on $\mathcal{X}$. Thus $u \in L^2(\mathcal{X})$, since P^{-1} is a bounded operator on $L^2(\mathcal{X})$. It is clear that u is a weak solution of the equation $Pu = f$, and hence, by regularity, u is a smooth solution. We only need to prove that $u \in V_{m,\varphi_1}^1(\mathcal{X}) \subseteq V_{m,\varphi_1}^{p_1}(\mathcal{X})$. Let us consider any g in $G_{\bar{\mathcal{F}},K} := \{g \in G \mid \mathrm{dist}\,(g\bar{\mathcal{F}}, K) > 1\}$. Since $\mathcal{X}$ is quasi-isometric to the metric space (G, d_S) via the orbit action by the Švarc-Milnor lemma, it is not hard to see that there are constants $C_1, C_2 > 0$ such that for every $g \in G_{\bar{\mathcal{F}},K}$ and every $(x,y) \in g\mathcal{F} \times K$, one has

$$2C_1|g| - C_2 \leq d_{\mathcal{X}}(x,y) \leq (2C_1)^{-1}|g| + C_2. \tag{3.63}$$

Taking $\delta = 1$, we can find $\varepsilon > 0$ so that the decay estimates (3.61) are satisfied. Now using Hölder's inequality, (3.61), and (3.63), we derive

$$\|u\|_{H^m(g\mathcal{F})} \lesssim \sup_{g\in G_{\bar{\mathcal{F}},K}} \max_{|\alpha|\leq m} \left(\int_{g\mathcal{F}} \left| \int_K |\partial_x^\alpha G_P(x,y)| \cdot |f(y)| \mathrm{d}\mu_{\mathcal{X}}(y) \right|^2 \mathrm{d}\mu_{\mathcal{X}}(x) \right)^{1/2}$$

$$\lesssim \|f\|_{L^2(\mathcal{X})} \cdot \max_{|\alpha|\leq m} \left(\int_{g\mathcal{F}} \int_K |\partial_x^\alpha G_P(x,y)|^2 \mathrm{d}\mu_{\mathcal{X}}(y) \mathrm{d}\mu_{\mathcal{X}}(x) \right)^{1/2}$$

$$\lesssim \|f\|_{L^2(\mathcal{X})} \cdot \exp(-2C_1\varepsilon|g|) \sup_{|\alpha|\leq m, y\in K} \left(\int_{g\mathcal{F}} |\partial_x^\alpha G_P(x,y)|^2 \exp(\varepsilon d_{\mathcal{X}}(x,y)) \mathrm{d}\mu_{\mathcal{X}}(x) \right)^{1/2}$$

$$\lesssim \|f\|_{L^2(\mathcal{X})} \cdot \exp(-2C_1\varepsilon|g|) \lesssim \|f\|_{L^2(\mathcal{X})} \cdot \varphi_1(g) \cdot \exp(-C_1\varepsilon|g|).$$

$$\tag{3.64}$$

Note that the above estimates hold up to multiplicative constants, which are uniform with respect to $g \in G_{\bar{\mathcal{F}},K}$. Therefore, u belongs to $V_{m,\varphi_1}^{p_1}(\mathcal{X})$, which proves (3.60).

In particular, the Fredholm indices of the operators $P_{m,\varphi_1}^{p_1}$ and $(P^*)_{m,\varphi_2}^{p_2}$ vanish. Now we are able to apply Theorem 1.26 to finish the proof of Lemma 3.1 in this case.

Case 2 $p_1^{-1} + p_2^{-1} \leq 1, \varphi_1^{-1}\varphi_2^{-1} \lesssim 1$. Consider a rigged divisor μ on $\mathcal{X}$. By assumptions, $L_{p_2'}(\mu, P, \varphi_2^{-1}) \subseteq L_{p_1}(\mu, P, \varphi_1)$ and

$$L_{p_1'}(\mu^{-1}, P^*, \varphi_1^{-1}) \subseteq L_{p_2}(\mu^{-1}, P^*, \varphi_2). \tag{3.65}$$

From these inclusions and Case 1, we get

$$\begin{aligned}
\dim L_{p_1}(\mu, P, \varphi_1) &= \deg_P(\mu) + \dim L_{p_1'}(\mu^{-1}, P^*, \varphi_1^{-1}) \\
&\leq \deg_P(\mu) + L_{p_2}(\mu^{-1}, P^*, \varphi_2) \\
&= \dim L_{p_2'}(\mu, P, \varphi_2^{-1}) \leq \dim L_{p_1}(\mu, P, \varphi_1).
\end{aligned} \tag{3.66}$$

Since all of the above inequalities must become equalities, this yields the corresponding Riemann-Roch formula in this case. **This finishes the proof of Lemma 3.1.**

We now use this lemma to prove all of the required statements.

First, one can get the identity in the second statement of Theorem 2.16 by taking $p_1 = p_2 = \infty$ and $\varphi_1 = \varphi_2 = \varphi_0$ in Lemma 3.1. Also, due to Theorem 2.14, there is no non-zero solution of P^* with subexponential growth. This implies that if $\mu^{-1} = (\emptyset, 0; D^+, L^+)$, the space $L_\infty(\mu^{-1}, P^*, \varphi_0)$ is trivial. Thus, the third statement follows immediately from the second statement. For the first statement, let us consider $p \in [1, \infty]$ and a function $\varphi \in \mathcal{S}(G)$. Now from Lemma 3.1 and the second statement, one gets:

- If φ is bounded,

$$\dim L_1(\mu, P, \varphi) = \deg_P(\mu) + \dim L_\infty(\mu^{-1}, P^*, \varphi_0) = \dim L_\infty(\mu, P, \varphi_0). \tag{3.67}$$

- If φ^{-1} is bounded and $1 \leq p \leq \infty$,

$$\dim L_p(\mu, P, \varphi) = \deg_P(\mu) + \dim L_\infty(\mu^{-1}, P^*, \varphi_0) = \dim L_\infty(\mu, P, \varphi_0). \tag{3.68}$$

We consider three cases.

Case 1. If φ is bounded, the two spaces $L_1(\mu, P, \varphi)$ and $L_\infty(\mu, P, \varphi_0)$ are the same since their dimensions are equal to each other by (3.67). Moreover,

$$L_1(\mu, P, \varphi) \subseteq L_p(\mu, P, \varphi) \subseteq L_p(\mu, P, \varphi_0) \subseteq L_\infty(\mu, P, \varphi_0). \tag{3.69}$$

This means that all these spaces are the same.

Case 2. If φ^{-1} is bounded, $L_\infty(\mu, P, \varphi_0) \subseteq L_\infty(\mu, P, \varphi)$. Using (3.68) with $p = \infty$, we have $L_\infty(\mu, P, \varphi_0) = L_\infty(\mu, P, \varphi)$. Moreover, (3.68) also yields that all the spaces $L_p(\mu, P, \varphi)$, where $1 \leq p \leq \infty$, must have the same dimension and therefore, they are the same space, which coincides with $L_\infty(\mu, P, \varphi_0)$.

Case 3. If neither φ nor φ^{-1} is bounded, we can consider the function $\phi := \varphi + \varphi^{-1}$. Clearly, ϕ is in $\mathcal{S}(G)$ and $\phi \geq 2$. Then according to Case 1 and Case 2,

$$L_p(\mu, P, \phi) = L_\infty(\mu, P, \varphi_0) = L_p(\mu, P, \phi^{-1}). \tag{3.70}$$

Also,

$$L_p(\mu, P, \phi^{-1}) \subseteq L_p(\mu, P, \varphi) \subseteq L_p(\mu, P, \phi), \tag{3.71}$$

since $\phi^{-1} \leq \varphi \leq \phi$. This means that $L_p(\mu, P, \varphi) = L_\infty(\mu, P, \varphi_0)$.

We can now prove the Corollary 2.17.

As in the proof of Corollary 2.9, the equivalence of the first three statements is an easy consequence of Theorem 2.16 and Remark 1.10. It is obvious that (iv) implies (iii). To see the converse, one can repeat the argument in the proof of Lemma 3.1 to show that the solution $u = P^{-1}f$ has exponential decay due to (3.61). By the unique solvability of the equation $Pu = f$ in $L^2(\mathcal{X})$, (iii) implies (iv).

Chapter 4
Specific Examples of Liouville-Riemann-Roch Theorems

Abstract In this chapter, we look at some examples of applications of the results of Chap. 2. These include in particular self-adjoint operators with non-degenerate spectral band edges, operators with Dirac points in dispersion relation, as well as some non-self-adjoint cases.

4.1 Self-Adjoint Operators

Let A be a bounded from below self-adjoint periodic elliptic operator of order m on an abelian co-compact covering X. We start with a brief reminder of some notions from Chap. 1.

For any real quasimomentum k, the operator $A(k)$ is self-adjoint, and its spectrum is discrete, consisting of real eigenvalues of finite multiplicities, which can be listed in non-decreasing order as

$$\lambda_1(k) \leq \lambda_2(k) \leq \ldots \nearrow \infty. \tag{4.1}$$

For each $j \in \mathbb{N}$, the function $k \mapsto \lambda_j(k)$ is called the jth **band function**. It is known (see e.g., [73]) that band functions λ_j are continuous, G^*-periodic and piecewise analytic in k. It is more convenient to consider the band functions as functions on the torus $\mathbb{T}^d$.

The range I_j of the jth band function is called the jth-**band**. The bands I_j can touch or overlap, but sometimes they may leave an open gap (Fig. 4.1). According to Theorem 1.4,

$$\sigma(A) = \bigcup_j I_j. \tag{4.2}$$

Therefore, the spectrum of a self-adjoint periodic elliptic operator A has a **band-gap structure**. An endpoint of a spectral gap is called a **gap edge** (or a **spectral edge**). To apply the results from Chap. 2, we will reformulate *Assumption* $\mathcal{A}$ from Chap. 2. For the relevant notations the reader is referred to Chap. 1.

© The Author(s), under exclusive license to Springer Nature Switzerland AG 2021
M. Kha, P. Kuchment, *Liouville-Riemann-Roch Theorems on Abelian Coverings*,
Lecture Notes in Mathematics 2245, https://doi.org/10.1007/978-3-030-67428-1_4

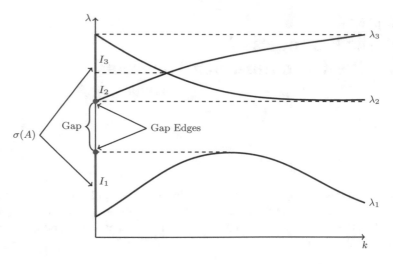

Fig. 4.1 An example of $\sigma(A)$

Assumption $\mathcal{A}'$
Suppose that the Fermi surface $F_{A,\mathbb{R}}$ is finite and consists of the points $\{k_1, \cdots, k_\ell\}$ (modulo G^*-shifts). Let

$$\{\lambda_{r,j}\}_{j=\overline{1,m_r}} \tag{4.3}$$

be the set of dispersion branches that are equal to 0 at the quasimomentum k_r ($1 \leq r \leq \ell$). There exists a family of pairwise disjoint neighborhoods V_r of k_r such that the function

$$k \in V_r \mapsto \max_{1 \leq j \leq m_r} |\lambda_{r,j}(k)|^{-1} \tag{4.4}$$

is L^1-integrable.[1]

Clearly, in the self-adjoint case the *Assumption $\mathcal{A}$* and *Assumption $\mathcal{A}'$* are equivalent.

Notation 4.1 *Given a natural number N,*

- *we denote by $h_{d,N}$ the dimension of the space of all harmonic polynomials of order at most N in d-variables, i.e.,*

$$h_{d,N} := \dim V_N^\infty(-\Delta_{\mathbb{R}^d}) = \binom{d+N}{d} - \binom{d+N-2}{d}; \tag{4.5}$$

---

[1] Note that for each $k \in V_r \setminus \{k_r\}$, $\lambda_{r,j}(k) \neq 0$ since $V_r \cap F_{A,\mathbb{R}} = \{k_r\}$.

- we also denote by $c_{d,N}$ the dimension of the space of all homogeneous polynomials of degree N in d variables, i.e.,

$$c_{d,N} := \binom{d+N}{d} - \binom{d+N-1}{d} = \binom{d+N-1}{N}. \tag{4.6}$$

4.1.1 Periodic Operators with Non-degenerate Spectral Edges

Let λ be an energy level that coincides with one of the gap edges in the spectrum of A. By shifting the spectrum, we can assume that $\lambda = 0$.

As it is discussed in Chap. 1, one can expect that the Fermi surfaces of A at the spectral edges normally are finite, and hence Liouville type results are applicable in these situations. We make the following assumption.

Assumption $\mathcal{B}$ There exists a band function $\lambda_j(k)$ such that for each quasimomentum k_r in the real Fermi surface $F_{A,\mathbb{R}}$, one has

($\mathcal{B}$1) 0 is a simple eigenvalue of the operator $A(k_r)$.
($\mathcal{B}$2) The Hessian matrix Hess $\lambda_j(k_0)$ is non-degenerate.

As it has been mentioned before, it is commonly believed in mathematics and physics literature (see e.g, [42]) that generically (with respect to the coefficients and other free parameters of the operator), extrema of band functions for second order operators of mathematical physics are isolated, attained by a single band and have non-degenerate Hessians, and thus satisfy the above assumption.

Suppose now that the free abelian rank d of the deck group G is greater than 2. The non-degeneracy assumption ($\mathcal{B}$2) implies the integrability of function $|\lambda_j(k)|^{-1}$ over a small neighborhood of $F_{A,\mathbb{R}}$. Hence, *Assumption $\mathcal{A}$* follows from *Assumption $\mathcal{B}$*.

Due to Theorem 1.17, the dimension of the space $V_N^\infty(A)$ is equal to $\ell h_{d,[N]}$ (see Notation 4.1). Applying the results in Chap. 2, we obtain the following results for a 'generic' self-adjoint second-order periodic elliptic operator A:

Theorem 4.2 *Suppose $d \geq 3$ and $N \in \mathbb{R}$. Let $\mu = (D^+, L^+; D^-, L^-)$ be a rigged divisor on X and, as before, $\mu^+ := (D^+, L^+; \emptyset, 0)$.*

a. If $N \geq 0$, then

$$\ell h_{d,[N]} + \deg_A(\mu) + \dim L_1(\mu^{-1}, A^*, -N) \leq \dim L_\infty(\mu, A, N)$$
$$\leq \ell h_{d,[N]} + \deg_A(\mu^+), \tag{4.7}$$

and

$$\dim L_\infty(\mu^+, A, N) = \ell h_{d,[N]} + \deg_A(\mu^+). \tag{4.8}$$

b. If $p \in [1, \infty)$ and $N > d/p$, then

$$\ell h_{d, \lfloor N-d/p \rfloor} + \deg_A(\mu) + \dim L_{p'}(\mu^{-1}, A^*, -N) \leq \dim L_p(\mu, A, N)$$
$$\leq \ell h_{d, \lfloor N-d/p \rfloor} + \deg_A(\mu^+),$$
(4.9)

and

$$\dim L_p(\mu^+, A, N) = \ell h_{d, \lfloor N-d/p \rfloor} + \deg_A(\mu^+).$$
(4.10)

c. For $d \geq 5$, the inequality

$$\deg_A(\mu) + \dim L_{p'}(\mu^{-1}, A^*, -N) \leq \dim L_p(\mu, A, N) \leq \deg_A(\mu^+) \quad (4.11)$$

holds, assuming one of the following two conditions:

- $p \in [1, 2)$, $N \in (d(2-p)/(2p), d/p]$.
- $p \in [2, \infty)$, $N \in [0, d/p]$.

Example 4.1

1. Let $A = A^* = -\Delta + V$ be a periodic Schrödinger operator with real-valued electric potential V on a co-compact abelian cover X. The domain of A is the Sobolev space $H^2(X)$, and thus, A is self-adjoint. Without loss of generality, we can suppose that 0 is the bottom of its spectrum. It is well-known [38] that *Assumption $\mathcal{B}$* holds in this situation. Hence, all the conclusions of Theorem 4.2 hold with $\ell = 1$, since $F_{A, \mathbb{R}} = \{0\}$ (modulo $2\pi \mathbb{Z}^d$-shifts).

2. In [14, 15], a deep analysis of the dispersion curves at the bottom of the spectrum was developed for a wide class of periodic elliptic operators of second-order on $\mathbb{R}^d$. Namely, let Γ be a lattice in $\mathbb{R}^d$ and Γ^* be its dual lattice, then these operators admit the following *regular factorization*:

$$A = \overline{f(x)} b(D)^* g(x) b(D) f(x),$$

where $b(D) = \sum_{j=1}^{d} -i \partial_{x_j} b_j : L^2(\mathbb{R}^d, \mathbb{C}^n) \rightarrow L^2(\mathbb{R}^d, \mathbb{C}^m)$ is a linear homogeneous differential operator whose coefficients b_j are constant $m \times n$-matrices of rank n (here $m \geq n$), f is a Γ-periodic and invertible $n \times n$ matrix function such that f and f^{-1} are in $L^\infty(\mathbb{R}^d)$, and g is a Γ-periodic and positive definite $m \times m$-matrix function such that for some constants $0 < c_0 \leq c_1$, $c_0 \mathbf{1}_{m \times m} \leq g(x) \leq c_1 \mathbf{1}_{m \times m}, x \in \mathbb{R}^d$. The existence of this factorization implies that the first band function attains a simple and nondegenerate minimum with value 0 at *only* the quasimomentum $k = 0$ (modulo Γ^*-shifts).

This covers the previous example since, it was noted from [14] that the periodic Schrödinger operator $D(g(x)D) + V(x)$ with a periodic metric $g(x)$

and a periodic potential $V(x)$ can be reformulated properly to admit a regular factorization.

3. Consider a self-adjoint periodic magnetic Schrödinger operator in $\mathbb{R}^n$ $(n > 2)$

$$H = (-i\nabla + A(x))^2 + V(x),$$

where $A(x)$ and $V(x)$ are real-valued periodic magnetic and electric potentials, respectively. Using a gauge transformation, we can always assume w.l.o.g. the following normalized condition on A:

$$\int_{\mathbb{T}^n} A(x)\mathrm{d}x = 0.$$

Note that the transpose of H is the magnetic Schrödinger operator

$$H^* = (-i\nabla - A(x))^2 + V(x).$$

From the discussion of [44] (see also [41, Theorem 3.1.7]), the lowest band function of H has a unique nondegenerate extremum at a single quasimomentum k_0 (modulo G^*-shifts) if the magnetic potential A is small enough, e.g., $\|A\|_{L^r(\mathbb{T}^n)} \ll 1$ or some $r > n$. Thus, we obtain the same conclusion as the case without magnetic potential. It is crucial that one has to assume the smallness of the magnetic potential since there are examples [65] showing that the bottom of the spectrum can be degenerate if the magnetic potential is large enough.

We end this part by providing an illustration of Corollary 2.12 for families of periodic elliptic operators with non-degenerate spectral edges.

Corollary 4.3 *Let A_0 be a periodic elliptic operator on a co-compact abelian covering X with the deck group $G = \mathbb{Z}^d$, where $d \geq 5$, such that its real Fermi surface $F_{A_0,\mathbb{R}}$ consists of finitely many simple and non-degenerate minima of the jth-band function of A_0 ($j \geq 1$). Let B be a symmetric and periodic differential operator on X such that B is A_0-bounded. We consider the perturbation $A_z = A_0 + zB$, $z \in \mathbb{R}$. By standard perturbation theory, there exists a continuous function $\lambda(z, k)$ defined for small z and all quasimomenta k such that $k \mapsto \lambda(z, k)$ is the jth band function of A_z and $\lambda_j(z, k)$ is analytic in z. Let λ_z be the minimum value of the band function $\lambda(z, k)$ of the perturbation A_z. Then for any rigged divisor μ, there exists $\varepsilon > 0$ such that*

$$\dim L_2(\mu, A_z - \lambda_z, 0) \leq \dim L_2(\mu, A_0, 0),$$

for any z satisfying $|z| < \varepsilon$.

The corresponding statement for non-degenerate maxima also holds.

4.1.2 Periodic Operators with Dirac Points

An important situation in solid state physics and material sciences is when two
branches of the dispersion relation touch, forming a conical junction point, which
is called a Dirac (or sometimes "diabolic") point. Two-dimensional massless Dirac
operators or $2D$-Schrödinger operators with honeycomb-symmetric potentials are
typical models of periodic operators with conical structures [12, 22, 32, 45].
Presence of such a point is the reason of the miraculous properties of graphene
and some other carbon allotropes.

Let us consider a self-adjoint periodic elliptic operator A such that there are
two branches λ_+ and λ_- of the dispersion relation of A that meet at $\lambda = 0, k =
k_r$, forming a Dirac cone. Equivalently, we can assume that locally around each
quasimomentum k_r in the Fermi surface $F_{A,\mathbb{R}}$, for some $c_r \neq 0$, one has

$$\lambda_+(k) = c_r |k - k_r| \cdot (1 + O(|k - k_r|)),$$

$$\lambda_-(k) = -c_r |k - k_r| \cdot (1 + O(|k - k_r|)).$$

It is immediate to see that the functions $|\lambda_+|^{-1}$ and $|\lambda_-|^{-1}$ are integrable over a
small neighborhood of $F_{A,\mathbb{R}}$ provided that $d > 1$. Hence, we conclude:

Theorem 4.4 *Suppose $d \geq 2$. Assume that the Fermi surface $F_{A,\mathbb{R}}$ consists of ℓ
Dirac conical points. Then, as in Theorem 4.2, we have*

a. *If $N \geq 0$, then*

$$2\ell c_{d,[N]} + \deg_A(\mu) + \dim L_1(\mu^{-1}, A^*, -N) \leq \dim L_\infty(\mu, A, N)$$

$$\leq 2\ell c_{d,[N]} + \deg_A(\mu^+)$$
$$(4.12)$$

and

$$\dim L_\infty(\mu^+, A, N) = 2\ell c_{d,[N]} + \deg_A(\mu^+). \qquad (4.13)$$

b. *If $p \in [1, \infty)$ and $N > d/p$, then*

$$2\ell c_{d,\lfloor N-d/p \rfloor} + \deg_A(\mu) + \dim L_{p'}(\mu^{-1}, A^*, -N) \leq \dim L_p(\mu, A, N)$$

$$\leq 2\ell c_{d,\lfloor N-d/p \rfloor} + \deg_A(\mu^+)$$
$$(4.14)$$

and

$$\dim L_p(\mu^+, A, N) = 2\ell c_{d,\lfloor N-d/p \rfloor} + \deg_A(\mu^+). \qquad (4.15)$$

c. *If $d \geq 3$ and a pair (p, N) satisfy the condition c. of Theorem 4.2 the conclusion
of Theorem 4.2 (c) also holds.*

Example 4.2 We consider here Schrödinger operators with honeycomb lattice potentials in $\mathbb{R}^2$.

Let us start with recalling briefly some notions from [12, 22]. The triangular lattice $\Lambda_h = \mathbb{Z}\mathbf{v}_1 \oplus \mathbb{Z}\mathbf{v}_2$ is spanned by the basis vectors:

$$\mathbf{v}_1 = a\left(\frac{\sqrt{3}}{2}, \frac{1}{2}\right)^t, \quad \mathbf{v}_2 = a\left(\frac{\sqrt{3}}{2}, -\frac{1}{2}\right)^t \ (a > 0). \tag{4.16}$$

The dual lattice is

$$\Lambda_h^* = \mathbb{Z}\mathbf{k}_1 \oplus \mathbb{Z}\mathbf{k}_2, \tag{4.17}$$

where

$$\mathbf{k}_1 = \frac{4\pi}{a\sqrt{3}}\left(\frac{1}{2}, \frac{\sqrt{3}}{2}\right)^t, \quad \mathbf{k}_2 = \frac{4\pi}{a\sqrt{3}}\left(\frac{1}{2}, -\frac{\sqrt{3}}{2}\right)^t. \tag{4.18}$$

We define

$$\mathbf{K} = \frac{1}{3}(\mathbf{k}_1 - \mathbf{k}_2), \quad \mathbf{K}' = -\mathbf{K}. \tag{4.19}$$

The Brillouin zone $\mathcal{B}_h$, a fundamental domain of the quotient $\mathbb{R}^2/\Lambda_h^*$, can be chosen as a hexagon in $\mathbb{R}^2$ such that the six vertices of this hexagon fall into two groups:

1. $\mathbf{K}$ type-vertices: $\mathbf{K}$, $\mathbf{K} + \mathbf{k}_2$, $\mathbf{K} - \mathbf{k}_1$.
2. $\mathbf{K}'$ type-vertices: $\mathbf{K}'$, $\mathbf{K}' - \mathbf{k}_2$, $\mathbf{K}' + \mathbf{k}_1$.

Note that these groups of vertices are invariant under the clockwise rotation $\mathcal{R}$ by $2\pi/3$.

A honeycomb lattice potential $V \in C^\infty(\mathbb{R}^2)$ is real, Λ_h-periodic, and there exists a point $x_0 \in \mathbb{R}^2$ such that V is inversion symmetric (i.e., even) and $\mathcal{R}$-invariant with respect to x_0 (see [22, Remark 2.4] and [12] for constructions of honeycomb lattice potentials). Now assume that V is a honeycomb lattice potential and consider the Schrödinger operator $H^\varepsilon = -\Delta + \varepsilon V$ ($\varepsilon \in \mathbb{R}$). One of the main results of [22] is that except possibly for ε in a countable and closed set $\tilde{C}$, the dispersion relation of H^ε has conical singularities at each vertex of $\mathcal{B}_h$. Assume that λ_j^ε, $j \in \mathbb{N}$, are the band functions of the operator H^ε for each $\varepsilon \in \mathbb{R}$. Then according to [22, Theorem 5.1], when $\varepsilon \notin \tilde{C}$, there exists some $j \in \mathbb{N}$ such that the Fermi surface $F_{H^\varepsilon, \lambda_j^\varepsilon(\mathbf{K})}$ of the operator H^ε at the level $\lambda_j^\varepsilon(\mathbf{K})$ contains (at least) two Dirac points located at the quasimomenta $\mathbf{K}$ and $\mathbf{K}'$ (modulo shifts by vectors in the dual lattice Λ_h^*). Now our next corollary is a direct consequence of [22, Theorem 5.1] and our previous discussion:

Corollary 4.5 *Let μ be a rigged divisor on $\mathbb{R}^2$ and V be a honeycomb lattice potential such that*

$$V_{1,1} := \int\limits_{\mathbb{R}^2/\Lambda_h} e^{-i(\mathbf{k}_1+\mathbf{k}_2)\cdot x} V(x)\mathrm{d}x \neq 0. \qquad (4.20)$$

Then for $\varepsilon \notin \tilde{C}$, there exists $j \in \mathbb{N}$ such that the following inequalities hold:

1. if $p = \infty, N \geq 0$:

$$\dim L_\infty(\mu, H^\varepsilon - \lambda_j^\varepsilon(\mathbf{K}), N) \geq 4([N]+1) + \deg_{H^\varepsilon - \lambda_j^\varepsilon(\mathbf{K})}(\mu)$$
$$+ \dim L_1(\mu^{-1}, H^\varepsilon - \lambda_j^\varepsilon(\mathbf{K}), -N), \qquad (4.21)$$

$$\dim L_\infty(\mu^+, H^\varepsilon - \lambda_j^\varepsilon(\mathbf{K}), N) \geq 4([N]+1) + \deg_{H^\varepsilon - \lambda_j^\varepsilon(\mathbf{K})}(\mu^+). \qquad (4.22)$$

2. If $1 \leq p < \infty, N > 2/p$:

$$\dim L_p(\mu, H^\varepsilon - \lambda_j^\varepsilon(\mathbf{K}), N) \geq 4(\lfloor N - 2/p \rfloor + 1) + \deg_{H^\varepsilon - \lambda_j^\varepsilon(\mathbf{K})}(\mu)$$
$$+ \dim L_{p'}(\mu^{-1}, H^\varepsilon - \lambda_j^\varepsilon(\mathbf{K}), -N), \qquad (4.23)$$

$$\dim L_p(\mu^+, H^\varepsilon - \lambda_j^\varepsilon(\mathbf{K}), N) \geq 4(\lfloor N - 2/p \rfloor + 1) + \deg_{H^\varepsilon - \lambda_j^\varepsilon(\mathbf{K})}(\mu^+). \qquad (4.24)$$

Moreover, there exists $\varepsilon_0 > 0$ such that for all $\varepsilon \in (-\varepsilon_0, \varepsilon_0) \setminus \{0\}$, we have

- *If $\varepsilon V_{1,1} > 0$, the above j can be chosen as $j = 1$.*
- *If $\varepsilon V_{1,1} < 0$, the above j can be chosen as $j = 2$.*

Remark 4.1 Other results on existence of Dirac points in the dispersion relations of Schrödinger operators with periodic potentials on honeycomb lattices are established for instance in [45] for quantum graph models of graphene and carbon nanotubes materials and in [12] for many interesting models including both discrete, quantum graph, and continuous ones.

4.2 Non-Self-Adjoint Second Order Elliptic Operators

We now consider a class of possibly non-self-adjoint second-order elliptic operators arising in probability theory. Let A be a G-periodic linear elliptic operator of second-order acting on functions u in $C^\infty(X)$ such that in local coordinate system $(U; x_1, \ldots, x_n)$, the operator A can be represented as

$$A = - \sum_{1 \le i,j \le n} a_{ij}(x) \partial_{x_i} \partial_{x_j} + \sum_{1 \le j \le n} b_j(x) \partial_{x_j} + c(x), \qquad (4.25)$$

where the coefficients a_{ij}, b_j, c are *real*, smooth, and G-periodic. The matrix $a(x) := (a_{ij}(x))_{1 \le i,j \le n}$ is positive definite. We notice that the coefficient $c(x)$ of zeroth-order of A is globally defined on X, since it is the image of the constant function 1 via A.

Definition 4.6 ([2, 51, 59])

a. A function u on X is called a G-**multiplicative** with exponent $\xi \in \mathbb{R}^d$, if it satisfies

$$u(g \cdot x) = e^{\xi \cdot g} u(x), \quad \forall x \in X, g \in G. \qquad (4.26)$$

In other words, u is a Bloch function with quasimomentum $i\xi$.
b. The *generalized principal eigenvalue* of A is defined by

$$\Lambda_A := \sup\{\lambda \in \mathbb{R} \mid (A - \lambda)u = 0 \text{ has a positive solution } u\}.$$

The principal eigenvalue is a generalized version of the bottom of the spectrum in the self-adjoint case (see e.g., [1]).

Let Λ^* be the formal adjoint operator to A. The generalized principal eigenvalues of A^* and A are equal, i.e., $\Lambda_A = \Lambda_{A^*}$. For an operator A of the type (4.25) and any $\xi \in \mathbb{R}^d$, it is known [2, 41, 51, 59] that there exists a unique real number $\Lambda_A(\xi)$ such that the equation $(A - \Lambda_A(\xi))u = 0$ has a *positive G-multiplicative* solution u. We list some known properties of this important function $\Lambda_A(\xi)$. The reader can find proofs in [2, 51, 59] (see also [44, Lemma 5.7]).

Proposition 4.7

a. $\Lambda_A = \max\limits_{\xi \in \mathbb{R}^d} \Lambda_A(\xi) = \Lambda_A(\xi_0)$ *for a unique ξ_0 in $\mathbb{R}^d$.*
b. *The function $\Lambda_A(\xi)$ is strictly concave, real analytic, bounded from above, and its gradient $\nabla \Lambda_A(\xi)$ vanishes at only its unique maximum point $\xi = \xi_0$. The Hessian of the function $\Lambda_A(\xi)$ is non-degenerate at all points.*
c. $\Lambda_A(\xi)$ *is the principal eigenvalue with multiplicity one of the operator $A(i\xi)$.*
d. $\Lambda_A \ge 0$ *if and only if A admits a positive periodic (super-) solution, which is also equivalent to the existence of a positive G-multiplicative solution u to the equation $Au = 0$.*
e. $\Lambda_A = 0$ *if and only if there is exactly one normalized positive solution u to the equation $Au = 0$.*

We are interested in studying Liouville-Riemann-Roch type results for such operators A satisfying $\Lambda_A(0) \geq 0$, which implies that A has a positive solution.

Example 4.3

1. Any operator A as in (4.25) without zeroth-order term satisfies $\Lambda_A = \Lambda_A(0) = 0$.
2. If the zeroth-order coefficient $c(x)$ of the operator A is nonnegative on X, $\Lambda_A(0)$ is also nonnegative.[2] Indeed, let u be a positive and periodic solution to the equation $Au = \Lambda_A(0)u$. If $\Lambda_A(0) < 0$, it follows from the equation that u is a positive and periodic subsolution, namely $Au < 0$ on X. By the strong maximum principle, u must be constant. This means that $0 \leq cu = Au < 0$, which is a contradiction!

Before stating the main result of this subsection, let us provide a key lemma.

Lemma 4.1

a. If $\Lambda_A(0) > 0$, then $F_{A,\mathbb{R}} = \emptyset$.
b. If $\Lambda_A(0) = 0$, then $F_{A,\mathbb{R}} = \{0\}$ *(modulo G^*-shifts). In this case, there exists an open strip V in $\mathbb{C}^d$ containing the imaginary axis $i\mathbb{R}^d$ such that for any $k \in V$, there is exactly one (isolated and nondegenerate eigenvalue) point $\lambda(k)$ in $\sigma(A(k))$ that is close to 0. The dispersion function $\lambda(k)$ is analytic in V and $\lambda(ik) = \Lambda_A(k)$ if $k \in \mathbb{R}^d$. Moreover,*

   - *When $\Lambda_A > 0$, $k = 0$ is a non-critical point of the dispersion $\lambda(k)$ in $V \cap \mathbb{R}^d$ as well as of the function $\Lambda_A(\cdot)$ in $\mathbb{R}^d$.*
   - *When $\Lambda_A = 0$, $k = 0$ is a non-degenerate extremum of the dispersion $\lambda(k)$ in $V \cap \mathbb{R}^d$ as well as of the function $\Lambda_A(\cdot)$ in $\mathbb{R}^d$.*

The statements of this lemma are direct consequences of [44, Lemma 5.8], Kato-Rellich theorem (see e.g., [61, Theorem XII.8]), and Proposition 4.7.

Theorem 4.8 *Let A be a periodic elliptic operator of second-order with real and smooth coefficients on X such that $\Lambda_A(0) \geq 0$. Let μ be a rigged divisor on X and μ^+ be its positive part. Then*

a. *If $\Lambda_A(0) > 0$,*

$$\dim L_\infty(\mu^+, A, \varphi) = \deg_A(\mu^+) \tag{4.27}$$

   and

$$\dim L_\infty(\mu, A, \varphi) = \deg_A(\mu) + \dim L_\infty(\mu^{-1}, A^*, \varphi^{-1}) \tag{4.28}$$

   for any function $\varphi \in \mathcal{S}(G)$ (see Definition 2.15).

---

[2]In general, the converse of this statement is not true: e.g., consider A^* in this case then the zeroth-order coefficient of the transpose A^* is not necessarily nonnegative while $\Lambda_{A^*}(0) = \Lambda_A(0) \geq 0$.

b. *If $\Lambda_A > \Lambda_A(0) = 0$ and $d \geq 2$, then*

- *For any $N \geq 0$,*

$$\dim L_\infty(\mu^+, A, N) = c_{d,[N]} + \deg_A(\mu^+) \qquad (4.29)$$

and

$$c_{d,[N]} + \deg_A(\mu) + \dim L_1(\mu^{-1}, A^*, -N) \leq \dim L_\infty(\mu, A, N)$$
$$\leq c_{d,[N]} + \deg_A(\mu^+). \qquad (4.30)$$

- *For any $p \in [1, \infty)$, $N > d/p$,*

$$\dim L_p(\mu^+, A, N) = c_{d,\lfloor N-d/p \rfloor} + \deg_A(\mu^+) \qquad (4.31)$$

and

$$c_{d,\lfloor N-d/p \rfloor} + \deg_A(\mu) + \dim L_{p'}(\mu^{-1}, A^*, -N) \leq \dim L_p(\mu, A, N)$$
$$\leq c_{d,\lfloor N-d/p \rfloor} + \deg_A(\mu^+). \qquad (4.32)$$

- *For $d \geq 3$ and a pair (p, N) satisfying the condition in Theorem 4.4 c.,*

$$\deg_A(\mu) + \dim L_{p'}(\mu^{-1}, A^*, -N) \leq \dim L_p(\mu, A, N) \leq \deg_A(\mu^+). \qquad (4.33)$$

c. *If $\Lambda_A = \Lambda_A(0) = 0$ and $d \geq 3$, then*

- *For any $N \geq 0$,*

$$\dim L_\infty(\mu^+, A, N) = h_{d,[N]} + \deg_A(\mu^+) \qquad (4.34)$$

and

$$h_{d,[N]} + \deg_A(\mu) + \dim L_1(\mu^{-1}, A^*, -N) \leq \dim L_\infty(\mu, A, N)$$
$$\leq h_{d,[N]} + \deg_A(\mu^+). \qquad (4.35)$$

- *For any $p \in [1, \infty)$, $N > d/p$,*

$$\dim L_p(\mu^+, A, N) = h_{d,\lfloor N-d/p \rfloor} + \deg_A(\mu^+) \qquad (4.36)$$

and

$$h_{d,\lfloor N-d/p \rfloor} + \deg_A(\mu) + \dim L_{p'}(\mu^{-1}, A^*, -N) \leq \dim L_p(\mu, A, N)$$

$$\leq h_{d,\lfloor N-d/p \rfloor} + \deg_A(\mu^+).$$

$$(4.37)$$

- *For $d \geq 5$ and a pair (p, N) satisfying the condition c. of Theorem 4.4,*

$$\deg_A(\mu) + \dim L_{p'}(\mu^{-1}, A^*, -N) \leq \dim L_p(\mu, A, N) \leq \deg_A(\mu^+).$$

$$(4.38)$$

Proof To compute the dimensions of the spaces $V_N^p(A)$, we use Lemma 4.1 to apply Theorem 1.17 and Theorem 1.13. Then the statements of Theorem 4.8 follows immediately from Theorem 2.16, Lemma 4.1, Theorem 2.2, Proposition 2.5, and Theorem 2.8.   □

Chapter 5
Auxiliary Statements and Proofs of Technical Lemmas

Abstract Here we collect a variety of technical auxiliary considerations and results used in, or related to the content of the main chapters of the book.

5.1 Properties of Floquet Functions on Abelian Coverings

We recall briefly another construction of Floquet functions on X. As we discussed in Sect. 1.4, it suffices to define the Bloch function $e_k(x)$ with quasimomentum k and the powers $[x]^j$ on X, where $j \in \mathbb{Z}_+^d$.

Definition 5.1 A smooth mapping h from X to $\mathbb{R}^d$ is called an *additive function* if the following condition holds:

$$h(g \cdot x) = h(x) + g, \tag{5.1}$$

where $(x, g) \in X \times \mathbb{Z}^d$.

There are various ways of constructing such additive functions (see e.g., [44, 51]). We will fix such an additive function h and write is as a tuple of scalar functions: $h = (h_1, \dots, h_d)$. Let $j = (j_1, \dots, j_d) \in \mathbb{Z}_+^d$ be a multi-index. We define

$$[x]^j := h(x)^j = \prod_{m=1}^{d} h_m(x)^{j_m}, \tag{5.2}$$

and

$$e_k(x) := \exp(ik \cdot h(x)). \tag{5.3}$$

Clearly, $e_k(g \cdot x) = e^{ik \cdot g} e_k(x)$. Then a Floquet function u of order N with quasimomentum k is of the form

$$u(x) = e_k(x) \sum_{|j| \leq N} p_j(x)[x]^j,$$

where p_j is smooth and periodic. Observe that the notion of Floquet functions is independent of the choice of h. Namely, u is also a Floquet function with the same order and quasimomentum with respect to another additive function $\tilde{h}$. Indeed, the difference $w := h - \tilde{h}$ between two additive functions h and $\tilde{h}$ is a periodic function. Hence, one can rewrite

$$u(x) = e^{ik \cdot \tilde{h}(x)} \sum_{|j| \leq N} e^{ik \cdot w(x)} p_j(x) \prod_{m=1}^{d} (\tilde{h}_m(x) + w_m(x))^{j_m}$$

$$= e^{ik \cdot \tilde{h}(x)} \sum_{|j| \leq N} e^{ik \cdot w(x)} p_j(x) \sum_{j' \leq j} \binom{j}{j'} w(x)^{j-j'} \tilde{h}(x)^{j'} \qquad (5.4)$$

$$= e^{ik \cdot \tilde{h}(x)} \sum_{|j| \leq N} \tilde{p}_j(x) \tilde{h}(x)^j,$$

where $\tilde{p}_j(x) := \sum_{j \leq j'} \binom{j'}{j} e^{ik \cdot w(x)} p_{j'}(x) w(x)^{j'-j}$ is periodic. The following simple lemma is needed later.

Lemma 5.1 *Let K be a compact neighborhood in X. Then for any multi-index $j \in \mathbb{Z}_+^d$, there exists some constant C such that for any $x \in K$ and $g \in \mathbb{Z}^d$, one has*

$$\left| [g \cdot x]^j - g^j \right| \leq C \langle g \rangle^{|j|-1}.$$

Proof

$$\left| [g \cdot x]^j - g^j \right| = \left| \prod_{m=1}^{d} (h_m(x) + g_m)^{j_m} - \prod_{m=1}^{d} g_m^{j_m} \right| \leq C \langle g \rangle^{|j|-1},$$

for some $C > 0$ depending on $\|h\|_{L^\infty(K)}$. $\square$

5.2 Basic Properties of the Family $\{A(k)\}_{k \in \mathbb{C}^d}$

We discuss another (equivalent) model of the operator family $A(k)$, which is sometimes useful.[1] In this part, we abuse notations and identify elements in $L^2(M)$ with their periodic extensions that belong to $L^2_{loc}(X)$.

Consider now an additive function h on the abelian covering X (see Definition 5.1). Let $\mathcal{U}_k$ be the mapping that multiplies a function $f(x)$ in $L^2_k(X)$ by $e^{-ik \cdot h(x)}$. Thus, $\mathcal{U}_k$ is an invertible bounded linear operator in $\mathcal{L}(L^2_k(X), L^2(M))$ and its inverse is given by the multiplication by $e^{ik \cdot h(x)}$. Note that the operator norms of $\mathcal{U}_k$ and $\mathcal{U}_k^{-1}$ are bounded by

$$e^{|\Im k| \cdot \|h\|_{L^\infty(\overline{\mathcal{F}})}}, \tag{5.5}$$

where $\mathcal{F}$ is a fundamental domain.

Consider the following elliptic operator:

$$\widehat{A}(k) := \mathcal{U}_k A(k) \mathcal{U}_k^{-1}. \tag{5.6}$$

The operator $\widehat{A}(k)$, with the Sobolev space $H^m(M)$ as the domain, is a closed and unbounded operator in $L^2(M)$. For each complex quasimomentum k, the two linear operators $\widehat{A}(k)$ and $A(k)$ are similar and thus, their spectra are identical. In terms of spectral information, it is no need to distinguish $A(k)$ and its equivalent model $\widehat{A}(k)$. One of the benefits of working with the later model is that $\widehat{A}(k)$ acts on the k-independent domain of periodic functions on X, while the differential expression becomes a polynomial in k. Moreover, the operator $A(k)$ now acts on sections of the appropriate linear bundle E_k (see Chap. 1).

The following proposition gives a simple sufficient condition on the principal symbol of the operator A so that the spectra of $A(k)$ are discrete. More general criteria on the discreteness of spectra can be found, for instance, in [3].

Proposition 5.2 *If A has real principal symbol, then for each $k \in \mathbb{C}^d$, $A(k)$, as an unbounded operator on $L^2(E_k)$, has discrete spectrum, i.e., its spectrum consists of isolated eigenvalues with finite (algebraic) multiplicities.*

Proof Let B be the real part of the operator A. Since A has real principal symbol, the principal symbols of A and B are the same. By pushing down to operators on M, the differential operator $\widehat{A}(0) - \widehat{B}(0)$ is of lower order. Also, the principal symbols of the operators $\widehat{A}(k)$ and $\widehat{A}(0)$ are identical. Thus, we see that $\widehat{A}(k)$ is a perturbation of the self-adjoint elliptic operator $\widehat{B}(0)$ by a lower order differential operator on the compact manifold M. It follows from [3] that the spectrum of $\widehat{A}(k)$ is discrete. This finishes the proof. □

---

[1]Compare with the discussion of Floquet multipliers in [42] and discussions of vector bundles E_k in Chap. 1.

If the spectrum $\sigma(A(k))$ is discrete, then the family of operators $\{\widehat{A}(k)\}_{k \in \mathbb{C}^d}$, has compact resolvents and is analytic of type (A) in the sense of Kato [33].[2] Therefore, this family satisfies the upper-semicontinuity of the spectrum (see e.g., [33, 61]). We provide this statement here without a proof.[3]

Proposition 5.3 *Consider $k_0 \in \mathbb{C}^d$. If Γ is a compact subset of the complex plane such that $\Gamma \cap \sigma(A(k_0)) = \emptyset$, then there exists $\delta > 0$ depending on Γ and k_0 such that $\Gamma \cap \sigma(A(k)) = \emptyset$, for any k in the ball $B_{k_0}(\delta)$ centered at k_0 with radius δ.*

Remark 5.1

(i) The Hilbert bundle $\mathcal{E}^m$ becomes the trivial bundle $\mathbb{C}^d \times H^m(M)$ via the holomorphic bundle isomorphism defined from the linear maps $\mathcal{U}_k$, where $k \in \mathbb{C}^d$.

(ii) In general, one can use the analytic Fredholm theorem to see that the essential spectrum[4] of $A(k)$ is empty for any $k \in \mathbb{C}^d$, but this is not enough to conclude that these spectra are discrete in the non-self-adjoint case. For example, if we consider the $\mathbb{Z}$-periodic elliptic operator $A = e^{2i\pi x} D_x$ on $\mathbb{R}$, a simple argument shows that

$$\sigma(A(k)) = \begin{cases} \mathbb{C}, & \text{if } k \in 2\pi\mathbb{Z} \\ \emptyset, & \text{otherwise.} \end{cases}$$

A similar example for the higher-dimensional case $\mathbb{R}^d$, $d > 1$ can be cooked up easily from the above example.

5.3   Properties of Floquet Transforms on Abelian Coverings

We describe here some useful properties of the Floquet transform $\mathbf{F}$ on abelian coverings (see more about this in [42]). First, due to (1.5), one can see that the Floquet transform $\mathbf{F}f(k, x)$ of a nice function f, e.g., $f \in C_c^\infty(X)$, is periodic in the quasimomentum variable k and moreover, it is a quasiperiodic function in the x-variable, i.e.,

$$\mathbf{F}f(k, g \cdot x) = \gamma_k(g) \cdot \mathbf{F}f(k, x) = e^{ik \cdot g} \cdot \mathcal{F}f(k, x), \quad \text{for any } (g, x) \in G \times X. \tag{5.7}$$

---

[2] A different approach to the analyticity of this operator family is taken in [41, 42].

[3] Stronger results about properties of spectra of analytic Fredholm operator functions are available in [76].

[4] Here we use the definition of the essential spectrum of an operator T as the set of all $\lambda \in \mathbb{C}$ such that $T - \lambda$ is not Fredholm.

It follows that $\mathbf{F}f(k, \cdot)$ belongs to $H_k^s(X)$, for any k and s. Therefore, it is enough to regard the Floquet transform of f as a smooth section of the Hilbert bundle $\mathcal{E}^s$ over the torus $\mathbb{T}^d$ (which can be identified with the Brillouin zone B).

Let $K \Subset X$ be a domain such that $\bigcup_{g \in G} gK = X$. Then given any real number s, we denote by $\mathcal{C}^s(X)$ the Frechet space consisting of all functions $\phi \in H_{loc}^s(X)$ such that for any $N \geq 0$, one has

$$\sup_{g \in G} \|\phi\|_{H^s(gK)} \cdot \langle g \rangle^N < \infty. \tag{5.8}$$

In terms of Definition 1.9,

$$\bigcap_{N \geq 0} V_N^\infty(X) = \mathcal{C}^0(X) \cap C^\infty(X). \tag{5.9}$$

The following theorem collects Plancherel type results for the Floquet transform.[5]

Theorem 5.4

(a) *The Floquet transform* $\mathbf{F}$ *is an isometric isomorphism between the Sobolev space* $H^s(X)$ *and the space* $L^2(\mathbb{T}^d, \mathcal{E}^s)$ *of* L^2*-integrable sections of the vector bundle* $\mathcal{E}^s$.

(b) *The Floquet transform* $\mathbf{F}$ *expands the periodic elliptic operator* A *of order* m *in* $L^2(X)$ *into a direct integral of the fiber operators* $A(k)$ *over* $\mathbb{T}^d$.

$$\mathbf{F}A\mathbf{F}^{-1} = \int_{\mathbb{T}^d}^\oplus A(k)\mathrm{d}k. \tag{5.10}$$

Equivalently, $\mathbf{F}(Af)(k) = A(k)\mathbf{F}f(k)$ *for any* $f \in H^m(X)$.

(c) *The Floquet transform*

$$\mathbf{F} : \mathcal{C}^s(X) \to C^\infty(\mathbb{T}^d, \mathcal{E}^s) \tag{5.11}$$

is a topological isomorphism, where $C^\infty(\mathbb{T}^d, \mathcal{E}^s)$ *is the space of smooth sections of the vector bundle* $\mathcal{E}^s$. *Furthermore, under the Floquet transform* $\mathbf{F}$, *the operator*

$$A : \mathcal{C}^m(X) \to \mathcal{C}^0(X) \tag{5.12}$$

becomes a morphism of sheaves of smooth sections arising from the holomorphic Fredholm morphism $A(k)$ *between the two holomorphic Hilbert bundles* $\mathcal{E}^m$ *and* $\mathcal{E}^0$ *over the torus* $\mathbb{T}^d$, *i.e., it is an operator from* $C^\infty(\mathbb{T}^d, \mathcal{E}^m)$ *to*

---

[5]Details, as well as Paley-Wiener type results can be found in [41–44].

$C^\infty(\mathbb{T}^d, \mathcal{E}^0)$ *such that it acts on the fiber of $\mathcal{E}^m$ at k as the fiber operator* $A(k) : H_k^m(X) \to L_k^2(X)$.

(d) The inversion $\mathbf{F}^{-1}$ of the Floquet transform is given by the formula

$$f(x) = \frac{1}{(2\pi)^d} \int_{\mathbb{T}^d} \mathbf{F} f(k, x) dk, \tag{5.13}$$

provided that one can make sense both sides of (5.13) (as functions or distributions).

We prove a simple analog of the Riemann-Lebesgue lemma for the Floquet transform.

Lemma 5.2

(a) Let $\widehat{f}(k, x)$ be a function in $L^1(\mathbb{T}^d, \mathcal{E}^0)$. Then the inverse Floquet transform $f := \mathbf{F}^{-1}\widehat{f}$ belongs to $L_{loc}^2(X)$ and

$$\sup_{g \in G} \|f\|_{L^2(g\mathcal{F})} < \infty.$$

Here $\mathcal{F}$ is a fixed fundamental domain. Moreover, one also has

$$\lim_{|g| \to \infty} \|f\|_{L^2(g\mathcal{F})} = 0.$$

(b) If $f \in V_0^1(X)$ then $\mathbf{F} f(k, x) \in C(\mathbb{T}^d, \mathcal{E}^0)$.

Proof We recall that $L - 2$ sections of $\mathcal{E}^0$ can be identified with the elements of $L^2(\mathcal{F})$.

To prove the first statement, we use the identity (5.13) and the Minkowski's inequality to obtain

$$\|f\|_{L^2(g\mathcal{F})} = \frac{1}{(2\pi)^d} \left\| \int_{\mathbb{T}^d} \mathbf{F} f(k, \cdot) dk \right\|_{L^2(g\mathcal{F})} = \frac{1}{(2\pi)^d} \left\| \int_{\mathbb{T}^d} e^{ik \cdot g} \mathbf{F} f(k, \cdot) dk \right\|_{L^2(\mathcal{F})}$$

$$\leq \frac{1}{(2\pi)^d} \int_{\mathbb{T}^d} \|\mathbf{F} f(k, \cdot)\|_{L^2(\mathcal{F})} dk = \frac{1}{(2\pi)^d} \|\mathbf{F} f(k, x)\|_{L^1(\mathbb{T}^d, \mathcal{E}^0)} < \infty. \tag{5.14}$$

To show that

$$\lim_{|g| \to \infty} \|f\|_{L^2(g\mathcal{F})} = 0, \tag{5.15}$$

one can easily modify the standard proof of the Riemann-Lebesgue lemma, i.e., by using Theorem 5.4 a. and then approximating $\mathbf{F} f$ by a sequence of functions in $L^2(\mathbb{T}^d, \mathcal{E}^0)$.

The second statement follows directly from (1.5) and the triangle inequality. □

5.4 A Schauder Type Estimate

For convenience, we state a well-known Schauder type estimate for solutions of a periodic elliptic operator A, which we need to refer to several times in this text. We also sketch its proof for the sake of completeness.

For any open subset $\mathcal{O}$ of X such that $\hat{K} \subset \mathcal{O}$, we define

$$G_{\hat{K}}^{\mathcal{O}} := \{ g \in G \mid g\hat{K} \subset \mathcal{O} \}. \tag{5.16}$$

Proposition 5.5 *Let $K \subset X$ be a compact set with non-empty interior and $\mathcal{O}$ be its open neighborhood. There exists a compact subset $\hat{K} \subset X$ such that $K \Subset \hat{K} \subset \mathcal{O}$ and the following statement holds: For any $\alpha \in \mathbb{R}^+$, there exists $C > 0$ depending on α, K, $\hat{K}$ such that*

$$\|u\|_{H^\alpha(gK)} \leq C \cdot \|u\|_{L^2(g\hat{K})}, \tag{5.17}$$

for any $g \in G_{\hat{K}}^{\mathcal{O}}$ and any solution $u \in C^\infty(\mathcal{O})$ satisfying the equation $Au = 0$ on $\mathcal{O}$.

Proof Let B be an almost local[6] pseudodifferential parametrix of A such that B commutes with actions of the deck group G (see e.g., [41, Lemma 2.1.1] or [67, Proposition 3.4]). Hence, $BA = 1+T$ for some almost-local and periodic smoothing operator T on X. This implies that for some compact neighborhood $\hat{K}$ (depending on the support of the Schwartz kernel of T and the subset K) and for any $\alpha \geq 0$, one can find some $C > 0$ so that for any smooth function v on a neighborhood of $\hat{K}$, one gets

$$\|Tv\|_{H^\alpha(K)} \leq C \cdot \|v\|_{L^2(\hat{K})}.$$

In particular, for any $g \in G_{\hat{K}}^{\mathcal{O}}$ and $u \in C^\infty(\mathcal{O})$,

$$\|Tu^g\|_{H^\alpha(K)} \leq C \cdot \|u^g\|_{L^2(\hat{K})}, \tag{5.18}$$

where u^g is the g-shift of the function u on $\mathcal{O}$. Since T is G-periodic, from (5.18), we obtain

$$\|Tu\|_{H^\alpha(gK)} \leq C \cdot \|u\|_{L^2(g\hat{K})}. \tag{5.19}$$

The important point here is the uniformity of the constant C with respect to $g \in G_{\hat{K}}^{\mathcal{O}}$.

---

[6]I.e., for some $\varepsilon > 0$, the support of the Schwartz kernel of B is contained in an ε-neighborhood of the diagonal of $X \times X$.

Suppose now that $Au = 0$ on $\mathcal{O}$. Thus, $u = BAu - Tu = -Tu$ on $\mathcal{O}$. This identity and (5.19) imply the estimate

$$\|u\|_{H^\alpha(gK)} = \|Tu\|_{H^\alpha(gK)} \leq C \cdot \|u\|_{L^2(g\hat{K})}, \quad \forall g \in G_{\hat{K}}^{\mathcal{O}}. \tag{5.20}$$

$\square$

Remark 5.2 We need to emphasize that Proposition 5.5 holds in a more general context. Namely, it is true for any C^∞-*bounded uniformly elliptic operator* P on a co-compact Riemannian covering $\mathcal{X}$ with a discrete deck group G. In this setting, P is invertible modulo a *uniform smoothing operator* T on $\mathcal{X}$ (see [67, Definition 3.1] and [67, Proposition 3.4]). Now the estimate (5.19) follows easily from the uniform boundedness of the derivatives of any order of the Schwartz kernel of T on canonical coordinate charts and a routine argument of partition of unity. Another possible approach is to invoke uniform local apriori estimates [67, Lemma 1.4].

5.5   A Variant of Dedekind's Lemma

It is a well-known theorem by Dedekind (see e.g., [55, Lemma 2.2]) that distinct unitary characters of an abelian group G are linearly independent as functions from G to a field $\mathbb{F}$. The next lemma is a refinement of Dedekind's lemma when $\mathbb{F} = \mathbb{C}$. We notice that a proof by induction method can be found in [60, Lemma 4.4]. For the sake of completeness, we will provide our analytic proof using Stone-Weierstrass's theorem.

Lemma 5.3 *Consider a finite number of distinct unitary characters* $\gamma_1, \ldots, \gamma_\ell$ *of the abelian group* $\mathbb{Z}^d$. *Then there are vectors* $g_1, \ldots, g_\ell$ *in* $\mathbb{Z}^d$ *and* $C > 0$ *such that for any* $v = (v_1, \ldots, v_\ell) \in \mathbb{C}^\ell$, *we have*

$$\max_{1 \leq s \leq \ell} \left| \sum_{r=1}^{\ell} v_r \cdot \gamma_r(g_s) \right| \geq C \cdot \max_{1 \leq r \leq \ell} |v_r|.$$

Proof By abuse of notation, we can regard $\gamma_1, \ldots, \gamma_\ell$ as distinct points of the torus $\mathbb{T}^d$.

For each tuple $(g_1, \ldots, g_\ell)$ in $(\mathbb{Z}^d)^\ell$, let $W(g_1, \ldots, g_\ell)$ be a $\ell \times \ell$-matrix whose (s, r)-entry $W(g_1, \ldots, g_\ell)_{s,r}$ is $\gamma_s^{g_r}$, for any $1 \leq r, s \leq \ell$. We equip $\mathbb{C}^\ell$ with the maximum norm. Then the conclusion of the lemma is equivalent to the invertibility of some operator $W(g_1, \ldots, g_\ell)$ acting from $\mathbb{C}^\ell$ to $\mathbb{C}^\ell$.

Suppose for contradiction that the determinant function $\det W(g_1, \ldots, g_\ell)$ is zero on $(\mathbb{Z}^d)^\ell$, i.e., for any $g_1, \ldots, g_\ell \in \mathbb{Z}^d$, one has

$$0 = \det W(g_1, \ldots, g_\ell) = \sum_{\sigma \in S_\ell} \text{sign}(\sigma) \cdot \left(\gamma_{\sigma(1)}^{g_1} \cdots \gamma_{\sigma(\ell)}^{g_\ell} \right),$$

where S_ℓ is the permutation group on $\{1, \dots, \ell\}$. Thus, the above relation also holds for any trigonometric polynomial $P(\gamma_1, \dots, \gamma_\ell)$ on $(\mathbb{T}^d)^\ell$, i.e.,

$$\sum_{\sigma \in S_\ell} \operatorname{sign}(\sigma) \cdot P(\gamma_{\sigma(1)}, \dots, \gamma_{\sigma(\ell)}) = 0.$$

By using the fact that the trigonometric polynomials are dense in $C((\mathbb{T}^d)^\ell)$ in the uniform topology (Stone-Weierstrass theorem), we conclude that

$$\sum_{\sigma \in S_\ell} \operatorname{sign}(\sigma) \cdot f(\gamma_{\sigma(1)}, \dots, \gamma_{\sigma(\ell)}) = 0, \qquad (5.21)$$

for any continuous function f on $(\mathbb{T}^d)^\ell$.

Now for each $1 \le r \le \ell$, let us select some smooth cutoff functions ω_r supported on a neighborhood of the point γ_r such that $\omega_r(\gamma_s) = 0$ whenever $s \ne r$. We define $f \in C((\mathbb{T}^d)^\ell)$ as follows

$$f(x_1, \dots, x_\ell) := \prod_{r=1}^{\ell} \omega_r(x_r), \quad x_1, \dots, x_\ell \in \mathbb{T}^d.$$

Hence, $f(\gamma_{\sigma(1)}, \dots, \gamma_{\sigma(\ell)})$ is non-zero if and only if σ is the trivial permutation. By substituting f into (5.21), we get a contradiction, which proves our lemma. □

5.6 Proofs of Some Other Technical Statements

In this section, we will use the notation $\mathcal{F}$ for the closure of a fundamental domain for G-action on the covering X.

5.6.1 Proof of Theorem 1.13

If $u \in V^\infty_{\lfloor N-d/p \rfloor}(A)$, then $u \in V^\infty_{N_0}(A)$ for some nonnegative integer N_0 such that $N_0 < N - d/p$. Thus,

$$\sum_{g \in G} \|u\|^p_{L^2(g\mathcal{F})} \langle g \rangle^{-pN} \lesssim \sum_{g \in G} \langle g \rangle^{p(N_0 - N)} < \infty. \qquad (5.22)$$

Hence, $V^\infty_{\lfloor N-d/p \rfloor}(A) \subseteq V^p_N(A)$.

Now suppose that $F_{A,\mathbb{R}} = \{k_1, \ldots, k_\ell\}$ (modulo G^*-shifts), where $\ell \in \mathbb{N}$. It suffices to show that

$$V_N^p(A) \subseteq V_{\lfloor N - d/p \rfloor}^\infty(A). \tag{5.23}$$

A key ingredient of the proof of (5.23) is the following statement.

Lemma 5.4 *Suppose that* $\mathcal{N} > d/p$.

(i) *If* $u \in V_{\mathcal{N}}^p(A) \cap V_{\mathcal{M}}^\infty(A)$ *for some* $0 \le \mathcal{M} < \mathcal{N} + 1 - d/p$, *then* $u \in V_{\mathcal{M}'}^\infty(A)$ *for some* $\mathcal{M}' < \mathcal{N} - d/p$. *In particular,* $u \in V_{\mathcal{N}-d/p}^\infty(A)$.

(ii) *If* u *is in* $V_{\mathcal{N}}^p(A) \cap V_{\mathcal{N}+1-d/p}^\infty(A)$, *then* $u \in V_{\mathcal{N}-d/p}^\infty(A)$.

Instead of proving Lemma 5.4 immediately, let us assume first its validity and prove (5.23). Consider $u \in V_N^p(A)$.

Case 1 $p > 1$.

We prove by induction that if $0 \le s \le d - 1$, then

$$u \in V_{N+d/p-(s+1)/p}^p(A) \cap V_{N-s/p}^\infty(A). \tag{5.24}$$

The statement holds for $s = 0$ since clearly, $V_N^p(A) \subseteq V_N^\infty(A)$ and $N+d/p-1/p \ge N$. Now suppose that (5.24) holds for s such that $s+1 \le d-1$. Since $1-1/p > 0$, we can apply Lemma 5.4 (i) to u and the pair $(\mathcal{N}, \mathcal{M}) = (N+d/p-(s+1)/p, N-s/p)$ to deduce that $u \in V_{N-(s+1)/p}^\infty(A)$. Therefore, (5.24) also holds for $s+1$. In the end, we have

$$u \in V_N^p(A) \cap V_{N-(d-1)/p}^\infty(A). \tag{5.25}$$

Applying Lemma 5.4 (i) again, we conclude that u belongs to $V_{\mathcal{M}'}^\infty(A)$ for some $\mathcal{M}' < N - d/p$. In other words, u is in $V_{\lfloor N-d/p \rfloor}^\infty(A)$.

Case 2 $p = 1$.

As in Case 1, we apply Lemma 5.4 (ii) and induction to prove that

$$u \in V_{N+d-1-s}^1(A) \cap V_{N-s}^\infty(A) \tag{5.26}$$

for any $0 \le s \le d - 1$. Hence,

$$u \in V_N^1(A) \cap V_{N+1-d}^\infty(A). \tag{5.27}$$

Due to Lemma 5.4 (ii) again, one concludes that

$$u \in V_N^1(A) \cap V_{N-d}^\infty(A). \tag{5.28}$$

Applying now Lemma 5.4 (i) to u and the pair $(\mathcal{N}, \mathcal{M}) = (N, N - d)$, we conclude that

$$u \in V^{\infty}_{\lfloor N-d \rfloor}(A) = V^{\infty}_{N-(d+1)}(A). \tag{5.29}$$

Thus, Theorem 1.13 follows from Lemma 5.4.

Let us turn now to our proof of Lemma 5.4, which consists of several steps.

(i) **Step 1.** The lemma is trivial if $u = 0$. So from now on, we assume that u is non-zero. Since $V^{p}_{\mathcal{N}}(A) \subseteq V^{\infty}_{\mathcal{N}}(A)$, we can apply Theorem 1.11 (ii) to represent $u \in V^{p}_{\mathcal{N}}(A)$ as a finite sum of Floquet solutions of A, i.e.,

$$u = \sum_{r=1}^{\ell} u_r. \tag{5.30}$$

Here u_r is a Floquet function of order $M_r \leq \mathcal{N}$ with quasimomentum k_r. Let N_0 be the highest order among all the orders of the Floquet functions u_r appearing in the above representation. Without loss of generality, we can assume that there exists $r_0 \in [1, \ell]$ such that for any $r \leq r_0$, the order M_r of u_r is maximal among all of these Floquet functions. Thus, $M_r = N_0 \leq \mathcal{M}$ when $r \in [1, r_0]$. To prove our lemma, it suffices to show that $N_0 < \mathcal{N} - d/p$.

Step 2. According to Proposition 5.5, we can pick a compact neighborhood $\hat{\mathcal{F}}$ of $\mathcal{F}$ such that for any $\alpha \geq 0$,

$$\|u\|_{H^{\alpha}(g\mathcal{F})} \leq C \cdot \|u\|_{L^2(g\hat{\mathcal{F}})} \tag{5.31}$$

for some $C > 0$ independent of $g \in G$.

Let $\alpha > n/2$, then the Sobolev embedding theorem yields the estimate

$$\|u\|_{C^0(g\mathcal{F})} \lesssim \|u\|_{L^2(g\hat{\mathcal{F}})}, \quad \forall g \in G. \tag{5.32}$$

From (5.32) and the fact that $u \in V^{p}_{\mathcal{N}}(A)$, we obtain

$$\sup_{x \in \mathcal{F}} \left(\sum_{g \in G} |u(g \cdot x)|^p \langle g \rangle^{-p\mathcal{N}} \right) \lesssim \sum_{g \in G} \|u\|^{p}_{L^2(g\hat{\mathcal{F}})} \langle g \rangle^{-p\mathcal{N}} < \infty. \tag{5.33}$$

Step 3. One can write

$$u(x) = \sum_{r=1}^{\ell} u_r(x) = \sum_{r=1}^{r_0} e_{k_r}(x) \sum_{|j|=N_0} a_{j,r}(x)[x]^j + O(|x|^{N_0-1}). \tag{5.34}$$

Here each function $a_{j,r}$ is G-periodic and the remainder term $O(|x|^{N_0-1})$ is an exponential-polynomial with periodic coefficients of order at most $N_0 - 1$. Hence, for any $(g, x) \in G \times \mathcal{F}$, we get

$$u(g \cdot x) = \sum_{r=1}^{r_0} e^{ik_r \cdot g} \sum_{|j|=N_0} e_{k_r}(x) a_{j,r}(x) [g \cdot x]^j + O(\langle g \rangle^{N_0-1}). \qquad (5.35)$$

Since $N_0 - 1 \le \mathcal{M} - 1 < \mathcal{N} - d/p$, the series

$$\sum_{g \in \mathbb{Z}^d} \langle g \rangle^{p(N_0-1)} \cdot \langle g \rangle^{-p\mathcal{N}} \qquad (5.36)$$

converges. From this and (5.33), we deduce that

$$\sup_{x \in \mathcal{F}} \sum_{g \in G} \left| \sum_{r=1}^{r_0} e^{ik_r \cdot g} \sum_{|j|=N_0} e_{k_r}(x) a_{j,r}(x) [g \cdot x]^j \right|^p \langle g \rangle^{-p\mathcal{N}} < \infty. \qquad (5.37)$$

By Lemma 5.1,

$$|[g \cdot x]^j - g^j| = O(\langle g \rangle^{N_0-1}) \qquad (5.38)$$

for any multi-index j such that $|j| = N_0$. This implies that

$$\sup_{x \in \mathcal{F}} \sum_{g \in G} \left| \sum_{|j|=N_0} \sum_{r=1}^{r_0} e_{k_r}(x) a_{j,r}(x) e^{ik_r \cdot g} g^j \right|^p \langle g \rangle^{-p\mathcal{N}} < \infty. \qquad (5.39)$$

Step 4. We will use Lemma 5.3 to reduce the condition (5.39) to the one without exponential terms $e^{ik_r \cdot g}$, so we could assume that $F_{A,\mathbb{R}} = \{0\}$ (modulo G^*-shifts).

Indeed, let $\gamma_1, \dots, \gamma_{r_0}$ be distinct unitary characters of G that are defined via $\gamma_r(g) := e^{ik_r \cdot g}$, where $r \in \{1, \dots, r_0\}$ and $g \in G$. Now, due to Lemma 5.3, there are $g_1, \dots, g_{r_0} \in G$ and a constant $C > 0$ such that for any vector $(v_1, \dots, v_{r_0}) \in \mathbb{C}^{r_0}$, we have the following inequality:

$$C \cdot \max_{1 \le s \le r_0} \left| \sum_{r=1}^{r_0} v_r \cdot e^{ik_r \cdot g_s} \right| \ge \max_{1 \le r \le r_0} |v_r|. \qquad (5.40)$$

Now, for any $(g, x) \in G \times \mathcal{F}$ and $1 \le s \le r_0$, we apply (5.40) to the vector

$$(v_1, \dots, v_{r_0}) := \left(\sum_{|j|=N_0} e_{k_r}(x) a_{j,r}(x)(g + g_s)^j \langle g + g_s \rangle^{-\mathcal{N}} e^{ik_r \cdot g} \right)_{1 \le r \le r_0} \qquad (5.41)$$

to deduce that

$$\max_{1 \le r \le r_0} \left| \sum_{|j|=N_0} e_{k_r}(x) a_{j,r}(x)(g+g_s)^j \right|^p \langle g+g_s \rangle^{-pN}$$

$$= \max_{1 \le r \le r_0} \left| \sum_{|j|=N_0} e_{k_r}(x) a_{j,r}(x)(g+g_s)^j \langle g+g_s \rangle^{-N} e^{ik_r \cdot g} \right|^p$$

$$\lesssim \max_{1 \le s \le r_0} \left| \sum_{r=1}^{r_0} \left(\sum_{|j|=N_0} e_{k_r}(x) a_{j,r}(x)(g+g_s)^j \langle g+g_s \rangle^{-N} e^{ik_r \cdot g} \right) e^{ik_r \cdot g_s} \right|^p$$

$$\lesssim \sum_{s=1}^{r_0} \left| \sum_{r=1}^{r_0} \sum_{|j|=N_0} e_{k_r}(x) a_{j,r}(x)(g+g_s)^j \cdot e^{ik_r \cdot (g+g_s)} \right|^p \cdot \langle g+g_s \rangle^{-pN}$$

$$(5.42)$$

Summing the estimate (5.42) over $g \in G$, we derive

$$\max_{1 \le r \le r_0} \sup_{x \in \mathcal{F}} \sum_{g \in G} \left| \sum_{|j|=N_0} e_{k_r}(x) a_{j,r}(x) g^j \right|^p \langle g \rangle^{-pN}$$

$$= \max_{1 \le r,s \le r_0} \sup_{x \in \mathcal{F}} \sum_{g \in G} \left| \sum_{|j|=N_0} e_{k_r}(x) a_{j,r}(x)(g+g_s)^j \right|^p \langle g+g_s \rangle^{-pN}$$

$$\lesssim \sup_{x \in \mathcal{F}} \sum_{s=1}^{r_0} \sum_{g \in G} \left| \sum_{r=1}^{r_0} \sum_{|j|=N_0} e_{k_r}(x) a_{j,r}(x)(g+g_s)^j \cdot e^{ik_r \cdot (g+g_s)} \right|^p \cdot \langle g+g_s \rangle^{-pN}$$

Therefore, we obtain

$$\max_{1 \le r \le r_0} \sup_{x \in \mathcal{F}} \sum_{g \in G} \left| \sum_{|j|=N_0} e_{k_r}(x) a_{j,r}(x) g^j \right|^p \langle g \rangle^{-pN}$$

$$(5.43)$$

$$\lesssim \sup_{x \in \mathcal{F}} \sum_{g \in G} \left| \sum_{r=1}^{r_0} \sum_{|j|=N_0} e_{k_r}(x) a_{j,r}(x) g^j \cdot e^{ik_r \cdot g} \right|^p \cdot \langle g \rangle^{-pN}.$$

From (5.39) and (5.43), we get

$$\sum_{g \in G} \left| \sum_{|j|=N_0} e_{k_r}(x) a_{j,r}(x) g^j \right|^p \cdot \langle g \rangle^{-pN} < \infty, \qquad (5.44)$$

for any $1 \le r \le r_0$ and $x \in \mathcal{F}$.

Step 5. At this step, we prove the following claim: If P is a non-zero homogeneous polynomial of degree N_0 in d-variables such that $N_0 < \mathcal{N} + 1 - d/p$ and

$$\sum_{g \in \mathbb{Z}^d} |P(g)|^p \cdot \langle g \rangle^{-p\mathcal{N}} < \infty, \tag{5.45}$$

then $N_0 < \mathcal{N} - d/p$.

Our idea is to approximate the series in (5.45) by the integral

$$\mathcal{I} := \int_{\mathbb{R}^d} |P(z)|^p \cdot \langle z \rangle^{-p\mathcal{N}} dz.$$

In fact, for any $z \in [0, 1)^d + g$, one can use the triangle inequality and the assumption that the order of P is N_0 to achieve the following estimate

$$|P(z)|^p \leq 2^{p-1} \left(|P(g)|^p + |P(z) - P(g)|^p \right) \lesssim |P(g)|^p + \langle g \rangle^{p(N_0-1)}.$$

Integrating the above estimate over the cube

$$[0, 1)^d + g \tag{5.46}$$

and then summing over all $g \in \mathbb{Z}^d$, we deduce

$$\mathcal{I} = \sum_{g \in \mathbb{Z}^d} \int_{[0,1)^d + g} |P(z)|^p \cdot \langle z \rangle^{-p\mathcal{N}} dz$$

$$\lesssim \sum_{g \in \mathbb{Z}^d} |P(g)|^p \cdot \langle g \rangle^{-p\mathcal{N}} + \sum_{g \in \mathbb{Z}^d} \langle g \rangle^{p(N_0-1-\mathcal{N})} < \infty, \tag{5.47}$$

where we have used (5.45) and the assumption $(N_0 - 1 - \mathcal{N})p < -d$.

We now rewrite the integral $\mathcal{I}$ in polar coordinates:

$$\mathcal{I} = \int_0^\infty \int_{\mathbb{S}^{d-1}} |P(r\omega)|^p \langle r \rangle^{-p\mathcal{N}} r^{d-1} d\omega dr$$

$$= \left(\int_0^\infty \langle r \rangle^{-p\mathcal{N}} r^{d-1+pN_0} dr \right) \cdot \int_{\mathbb{S}^{d-1}} |P(\omega)|^p d\omega. \tag{5.48}$$

Suppose for contradiction that $(N_0 - \mathcal{N})p \geq -d$. Then

$$\int_0^\infty \langle r \rangle^{-p\mathcal{N}} r^{d-1+pN_0} \mathrm{d}r = \infty. \tag{5.49}$$

Thus, the finiteness of $\mathcal{I}$ implies that

$$\int_{\mathbb{S}^{d-1}} |P(\omega)|^p \mathrm{d}\omega = 0. \tag{5.50}$$

Hence, $P(\omega) = 0$ for any $\omega \in \mathbb{S}^{d-1}$. By homogeneity, P must be zero, which is a contradiction that proves our claim.

Step 6. Since u is non-zero, there are some $r \in \{1, \ldots, r_0\}$ and $x \in \mathcal{F}$ such that the following homogeneous polynomial of degree N_0 in $\mathbb{R}^d$

$$P(z) := \sum_{|j|=N_0} e_{k_r}(x) a_{j,r}(x) z^j \tag{5.51}$$

is non-zero. Due to (5.44) and the condition

$$N_0 \leq \mathcal{M} < \mathcal{N} + 1 - d/p \text{ (see Step 1)}, \tag{5.52}$$

the inequality $N_0 < \mathcal{N} - d/p$ must be satisfied according to Step 5. This finishes the proof of the first part of the lemma.

(ii) Consider

$$u \in V_{\mathcal{N}}^p(A) \cap V_{\mathcal{N}+1-d/p}^\infty(A). \tag{5.53}$$

In particular, for any $\varepsilon > 0$,

$$u \in V_{\mathcal{N}+\varepsilon}^p(A) \cap V_{\mathcal{N}+\varepsilon+1-d/p}^\infty(A). \tag{5.54}$$

We repeat the proof of Lemma 5.4 (i) for $(\mathcal{N} + \varepsilon)$ (instead of $\mathcal{N}$ in part (i)). As in the Step 1 of the previous proof, we first decompose u as a finite sum of Floquet solutions and let N_0 be the highest order among all the orders of the Floquet functions appearing in that decomposition. Repeating all the steps of the part (i), we conclude that

$$N_0 < \mathcal{N} + \varepsilon - d/p \tag{5.55}$$

for any $\varepsilon > 0$. By letting $\varepsilon \to 0^+$, $N_0 \leq \mathcal{N} - d/p$. We conclude that $u \in V_{\mathcal{N}-d/p}^\infty(A)$. This yields the second part of the lemma.

5.6.2 Proof of Theorem 1.14

(a) Consider $u \in V_N^p(A)$. Due to Theorem 1.13 and the condition that $N \leq d/p$,

$$V_N^p(A) \subseteq V_{d/p+1/2}^p(A) = V_0^\infty(A). \tag{5.56}$$

Using Theorem 1.11 (ii), we get

$$u(x) = \sum_{r=1}^{\ell} e_{k_r}(x)a_r(x), \tag{5.57}$$

for some periodic functions $a_r(x)$.

Using (5.32) and the assumption that $u \in V_N^p(A)$, we derive

$$\sup_{x \in \mathcal{F}} \sum_{g \in G} \left| \sum_{r=1}^{\ell} e_{k_r}(x)a_r(x)e^{ik_r \cdot g} \right|^p \cdot \langle g \rangle^{-pN} < \infty. \tag{5.58}$$

Now one can modify (from the estimate (5.44) instead of (5.39)) the argument in Step 4 of the proof of Lemma 5.4 to get

$$\max_{1 \leq r \leq \ell} \sup_{x \in \mathcal{F}} \left| e_{k_r}(x)a_r(x) \right|^p \cdot \sum_{g \in \mathbb{Z}^d} \langle g \rangle^{-pN} < \infty. \tag{5.59}$$

Hence, the assumption $-pN \geq -d$ implies that $\max_{1 \leq r \leq \ell} \sup_{x \in \mathcal{F}} \left| e_{k_r}(x)a_r(x) \right| = 0$.

Thus, u must be zero.

(b) Let u be an arbitrary element in $V_N^\infty(A)$. Since $N < 0$, we can assume that u has the form (5.57). To prove that $u = 0$, it is enough to show that $e_{k_r}(x)a_r(x) = 0$ for any $x \in \mathcal{F}$ and $1 \leq r \leq \ell$. One can repeat the same argument of the previous part to prove this claim. However, we will provide a different proof using Fourier analysis on the torus $\mathbb{T}^d$.

For each $x \in \mathcal{F}$, we introduce the following distribution on $\mathbb{T}^d$

$$f(k) := \sum_{r=1}^{\ell} e_{k_r}(x)a_r(x)\delta(k - k_r), \tag{5.60}$$

where $\delta(\cdot - k_r)$ is the Dirac delta distribution on the torus $\mathbb{T}^d$ at the quasimomentum k_r. In terms of Fourier series, we obtain

$$\hat{f}(g) = \sum_{r=1}^{\ell} e_{k_r}(x)a_r(x)e^{-ik_r \cdot g}. \tag{5.61}$$

As in (5.58), the assumption $u \in V_N^\infty(A)$ is equivalent to

$$\sup_{g \in \mathbb{Z}^d} \left| \hat{f}(g) \right| \cdot \langle g \rangle^{-N} < \infty. \tag{5.62}$$

Let ϕ be a smooth function on $\mathbb{T}^d$. Using Parseval's identity and Hölder's inequality, we obtain

$$\left| \sum_{r=1}^{\ell} e_{k_r}(x) a_r(x) \phi(k_r) \right| = |\langle f, \phi \rangle| = \left| \sum_{g \in \mathbb{Z}^d} \hat{f}(g) \hat{\phi}(-g) \right| \lesssim \sum_{g \in \mathbb{Z}^d} |\hat{\phi}(g)| \cdot \langle g \rangle^N. \tag{5.63}$$

Let us now pick $\delta > 0$ small enough such that $k_s \notin B(k_r, 2\delta)$ if $s \neq r$. Then we choose a cut-off function ϕ_r such that $\mathrm{supp}\,\phi_r \subseteq B(k_r, 2\delta)$ and $\phi_r = 1$ on $B(k_r, \delta)$. For $1 \le r \le \ell$, we define functions in $C^\infty(\mathbb{T}^d)$ as follows:

$$\phi_r^\varepsilon(k) := \phi_r(\varepsilon^{-1}k), \quad (0 < \varepsilon < 1). \tag{5.64}$$

To bound the Fourier coefficients of ϕ_r^ε in terms of ε, we use integration by parts. Indeed, for any nonnegative real number s,

$$
\begin{aligned}
(2\pi)^d |\widehat{\phi_r^\varepsilon}(g)| &= \left| \int_{\mathbb{T}^d} \phi_r^\varepsilon(k) e^{-ik \cdot g} dk \right| = \varepsilon^d \cdot \left| \int_{B(k_r, 2\delta)} \phi_r(k) e^{-i\varepsilon k \cdot g} dk \right| \\
&= \varepsilon^d \langle \varepsilon g \rangle^{-2[s]-2} \cdot \left| \int_{B(k_r, 2\delta)} (1 - \Delta)^{[s]+1} \phi_r(k) \cdot e^{-ik \cdot \varepsilon g} dk \right| \\
&\le \varepsilon^d \langle \varepsilon g \rangle^{-2[s]-2} \cdot \sup_k \left| (1 - \Delta)^{[s]+1} \phi_r(k) \right| \lesssim \varepsilon^{d-s} \langle g \rangle^{-s}.
\end{aligned}
\tag{5.65}
$$

In the last inequality, we make use of the fact that

$$\langle \varepsilon g \rangle^{-2[s]-2} \le \langle \varepsilon g \rangle^{-s} \le \varepsilon^{-s} \langle g \rangle^{-s} \tag{5.66}$$

whenever $\varepsilon \in (0, 1)$. In particular, by choosing any $s \in (\max(0, N+d), d)$, one has

$$|\widehat{\phi_r^\varepsilon}(g)| \cdot \langle g \rangle^N \lesssim \varepsilon^{d-s} \cdot \langle g \rangle^{N-s}. \tag{5.67}$$

Now we substitute $\phi := \phi_r^\varepsilon$ in (5.63), use (5.67), and then take $\varepsilon \to 0^+$ to derive

$$|e_{k_r}(x) a_r(x)| \lesssim \lim_{\varepsilon \to 0} \varepsilon^{d-s} \sum_{g \in \mathbb{Z}^d} \langle g \rangle^{N-s} = 0. \tag{5.68}$$

5.6.3 *Proof of Corollary 1.27*

It suffices to prove that $\operatorname{Im} \tilde{P} = \tilde{\Gamma}_\mu(\mathcal{X}, P)$, since according to (1.35),

$$\dim \operatorname{Ker} \tilde{P}^* = \operatorname{codim} \operatorname{Im} \tilde{P} = 0 \tag{5.69}$$

and then the conclusion of Corollary 1.27 holds true as we mentioned in Sect. 1.8. Now, given any $f \in \tilde{\Gamma}_\mu(\mathcal{X}, P)$, one has $\langle f, \tilde{L}^- \rangle = 0$ and $f = Pu$ for some $u \in \operatorname{Dom} P$, since $\operatorname{Im} P = \operatorname{Dom}' P^*$. According to the assumption, we can find a solution $w = u - v$ in $\operatorname{Dom} P$ of the equation $Pw = Pu - Pv = f$ such that $\langle w, L^- \rangle = 0$. Let w_0 be the restriction of w on $\mathcal{X} \setminus D^+$. Clearly, w_0 belongs to the space $\Gamma(\mathcal{X}, \mu, P)$. Since Pw is smooth on $\mathcal{X}$, $\tilde{P} w_0 = Pw = f$ by the definition of the extension operator $\tilde{P}$. This shows that $f \in \operatorname{Im} \tilde{P}$, which finishes the proof.

Remark 5.3 In the special case when $D^- = \emptyset$, $L^- = \{0\}$, one can prove the Riemann-Roch equality (1.29) directly, i.e., without referring to the extension operators $\tilde{P}$ and $\tilde{P}^*$. For reader's convenience, we present this short proof following [30]. We define the space $\Gamma(\mu, P) := \{u \in \mathcal{D}'(\mathcal{X}) \mid u \in \operatorname{Dom}_{D^+} P, \, Pu \in L^+\}$. Then it is easy to check that the following sequences are exact:

$$0 \to \tilde{L}^+ \xrightarrow{i} \Gamma(\mu, P) \xrightarrow{r} L(\mu, P) \to 0$$

$$0 \to \operatorname{Ker} P \xrightarrow{i} \Gamma(\mu, P) \xrightarrow{P} L^+ \to 0,$$

where i and r are natural inclusion and restriction maps. Here the surjectivity of P from $\Gamma(\mu, P)$ to L^+ is a consequence of the existence of a properly supported pseudodifferential parametrix of P (modulo a properly supported smoothing operator) and $C_c^\infty(\mathcal{X}) \subseteq \operatorname{Dom}' P^* = \operatorname{Im} P$. Note that $\operatorname{Ker} P^* = \{0\}$. Hence, it follows that

$$\dim L(\mu, P) = \dim \Gamma(\mu, P) - \dim \tilde{L}^+ = \dim \operatorname{Ker} P + \dim L^+ - \dim \tilde{L}^+$$

$$= \operatorname{ind} P + \deg_P(\mu).$$

Appendix A
Final Remarks and Conclusions

- An interesting discussion of issues related to the difference operators Definition 1.7 of Floquet functions can be found in the recent papers [35, 36]. There, a study of polynomial-like elements in vector spaces equipped with group actions is provided. These elements are defined via iterated difference operators. In the case of a full rank lattice acting on an Euclidean space, they are exactly polynomials with periodic coefficients, and thus are closely related to solutions of periodic differential equations. The main theorem of that work confirms that if the space of polynomial-like elements of degree zero is of finite dimension then for any $n \in \mathbb{Z}_+$, the space consisting of all polynomial-like elements of degree at most n is also finite dimensional. Non-abelian groups are considered, with the hope to transfer at least some of the Liouville theorems to the case of nilpotent co-compact coverings (compare with [51]), albeit this goal has not been achieved yet.

- The Remark 3.1 shows that Assumption ($\mathcal{A}2$) cannot be dropped in Theorem 2.2. Besides the example given in Remark 3.1, we provide a heuristic explanation here. It is known (see, e.g. [76]) that if $\{A_t\}$ is a family of Fredholm operators that is continuous with respect to the parameter t, the kernel dimension $\dim \operatorname{Ker} A_t$ is upper semicontinuous. The idea in combining Riemann-Roch and Liouville theorems by considering dimensions of spaces of solutions with polynomial growth as some Fredholm indices would imply that the upper-semicontinuity property should hold also for these dimensions (Corollary 2.12). On the other hand, as shown in [65], there exists a continuous family $\{M_t\}$ of periodic operators on $\mathbb{R}^2$ such that for each $N \geq 0$,

$$\dim V_N^\infty(M_t) = 2 \dim V_N^\infty(M_{2\sqrt{3}}) \tag{A.1}$$

if $2\sqrt{3} < t < 2\sqrt{3} + \varepsilon$ for some $\varepsilon > 0$ and thus, $\dim V_N(M_t)$ is not upper-semicontinuous at $t = 2\sqrt{3}$ (see [44]). In this example, the minimum 0 of the lowest band $\lambda_1(k)$ of the operator $M_{2\sqrt{3}}$ is degenerate [65]. This explains why our

© The Author(s), under exclusive license to Springer Nature Switzerland AG 2021
M. Kha, P. Kuchment, *Liouville-Riemann-Roch Theorems on Abelian Coverings*,
Lecture Notes in Mathematics 2245, https://doi.org/10.1007/978-3-030-67428-1

approach requires the "non-degeneracy" type condition ($\mathcal{A}2$) for avoiding some intractable cases like the previous example. Notice that in general, Assumption $\mathcal{A}$ and Liouville type results are not stable under small perturbations.

- Since both the Riemann-Roch type results of [30, 31] and the Liouville type results of [44] hold for elliptic systems, the results of this work could be easily extended (at the expense of heavier notations) to linear elliptic matrix operators (e.g., of Maxwell type) acting between vector bundles.
- Our results do not cover the important case when $A = \bar{\partial}$ on an abelian covering of a compact complex manifold. We plan to study this case in a separate paper.
- A natural question that arises is of having the divisor being also periodic, rather than compact (compare with [68]), which would require measuring "dimensions" of some infinite dimensional spaces, which in turn would require using some special von Neumann algebras and the corresponding traces. This task does not seem to be too daunting, but the authors have not addressed it here.
- This work shows that adding a pole at infinity (Liouville theorems) to the Riemann-Roch type results is not automatic and not always works. Moreover, the choice of the L_p space for measuring growth in Liouville theorems is very relevant for the results (in [43, 44] only $p = \infty$ was considered). In particular, when $p = 2$, the Liouville-Riemann-Roch equality we obtained in (2.7) could be viewed as an analog of the L^2-Riemann-Roch theorem in [68] for finite divisors.

References

1. S. Agmon, *On Positivity and Decay of Solutions of Second Order Elliptic Equations on Riemannian Manifolds* (Liguori, Naples, 1983)
2. S. Agmon, *On Positive Solutions of Elliptic Equations with Periodic Coefficients in* $\mathbf{R}^n$, *Spectral Results and Extensions to Elliptic Operators on Riemannian Manifolds*. Differential Equations (Birmingham, Alabama, 1983), 1984, pp. 7–17
3. S. Agmon, *Lectures on Elliptic Boundary Value Problems* (AMS Chelsea Publishing, Providence, RI, 2010)
4. N.W. Ashcroft, N.D. Mermin, *Solid State Physics* (Holt, Rinehart and Winston, New York, London, 1976)
5. M.F. Atiyah, *Elliptic operators, discrete groups and von Neumann algebras*, Colloque "Analyse et Topologie" en l'Honneur de Henri Cartan (Orsay, 1974), 1976, pp. 43–72. Astérisque, No. 32–33
6. M.F. Atiyah, I.M. Singer, The index of elliptic operators. I. Ann. Math. (2) **87**, 484–530 (1968). https://doi.org/10.2307/1970715
7. M.F. Atiyah, I.M. Singer, The index of elliptic operators. III. Ann. Math. (2) **87**, 546–604 (1968). https://doi.org/10.2307/1970717
8. M.F. Atiyah, I.M. Singer, The index of elliptic operators. IV. Ann. Math. (2) **93**, 119–138 (1971). https://doi.org/10.2307/1970756
9. M.F. Atiyah, I.M. Singer, The index of elliptic operators. V. Ann. Math. (2) **93**, 139–149 (1971). https://doi.org/10.2307/1970757
10. M.F. Atiyah, I.M. Singer, The index of elliptic operators on compact manifolds. Bull. Am. Math. Soc. **69**, 422–433 (1963). https://doi.org/10.1090/S0002-9904-1963-10957-X
11. M. Avellaneda, F.-H. Lin, Un théorème de Liouville pour des équations elliptiques à coefficients périodiques. C. R. Acad. Sci. Paris Sér. I Math. **309**(5), 245–250 (1989)
12. G. Berkolaiko, A. Comech, Symmetry and Dirac points in graphene spectrum. J. Spectr. Theory **8**(3), 1099–1147 (2018). https://doi.org/10.4171/JST/223
13. G. Berkolaiko, P. Kuchment, *Introduction to Quantum Graphs* (AMS, Providence, RI, 2013)
14. M. Birman, T. Suslina, *Threshold Effects Near the Lower Edge of the Spectrum for Periodic Differential Operators of Mathematical Physics*. Systems, Approximation, Singular Integral Operators, and Related Topics (Bordeaux, 2000), 2001, pp. 71–107
15. M. Birman, T. Suslina, Periodic second-order differential operators. Threshold properties and averaging. Algebra i Analiz **15**(5), 1–108 (2003)
16. T.H. Colding, W.P. Minicozzi II, Harmonic functions on manifolds. Ann. Math. **146**(3), 725–747 (1997)

17. T.H. Colding, W.P. Minicozzi I, An excursion into geometric analysis. Surv. Differ. Geometry **IX** 83–146 (2004). https://doi.org/10.4310/SDG.2004.v9.n1.a4
18. T.H. Colding, W.P. Minicozzi I, Liouville properties. ICCM Not. **7**(1), 16–26 (2019). https://doi.org/10.4310/ICCM.2019.v7.n1.a10
19. H.L. Cycon, R.G. Froese, B. Simon, *Schrödinger Operators with Application to Quantum Mechanics and Global Geometry*. Study, Texts and Monographs in Physics (Springer, Berlin, 1987)
20. M.S.P. Eastham, *The Spectral Theory of Periodic Differential Equations*. Texts in Mathematics (Edinburgh) (Scottish Academic Press, Edinburgh; Hafner Press, New York, 1973)
21. J. Eichhorn, *Global Analysis on Open Manifolds* (Nova Science Publishers, New York, 2007)
22. C.L. Fefferman, M.I. Weinstein, Honeycomb lattice potentials and Dirac points. J. Am. Math. Soc. **25**(4), 1169–1220 (2012)
23. J. Feldman, H. Knörrer, E. Trubowitz, Asymmetric Fermi surfaces for magnetic Schrödinger operators. Comm. Partial Differ. Equ. **25**(1–2), 319–336 (2000)
24. I.M. Gel′fand, Expansion in characteristic functions of an equation with periodic coefficients. Doklady Akad. Nauk SSSR (N.S.) **73**, 1117–1120 (1950)
25. I.M. Gel′fand, On elliptic equations. Russ. Math. Surv. **15**(3), 113–123 (1960). https://doi.org/10.1070/RM1960v015n03ABEH004094
26. I.M. Glazman, *Direct Methods of Qualitative Spectral Analysis of Singular Differential Operators*. Translated from the Russian by the IPST staff, Israel Program for Scientific Translations, Jerusalem, 1965 (Daniel Davey & Co., New York, 1966)
27. R.I. Grigorchuk, Degrees of growth of finitely generated groups and the theory of invariant means. Izv. Akad. Nauk SSSR Ser. Mat. **48**(5), 939–985 (1984)
28. M. Gromov, Groups of polynomial growth and expanding maps. Inst. Hautes Études Sci. Publ. Math. **53**, 53–73 (1981)
29. M. Gromov, M.A. Shubin, The Riemann-Roch theorem for general elliptic operators. C. R. Acad. Sci. Paris Sér. I Math. **314**(5), 363–367 (1992)
30. M. Gromov, M.A. Shubin, The Riemann-Roch theorem for elliptic operators. I. M. Gel′fand Seminar, 211–241 (1993)
31. M. Gromov, M.A. Shubin, The Riemann-Roch theorem for elliptic operators and solvability of elliptic equations with additional conditions on compact subsets. Invent. Math. **117**(1), 165–180 (1994)
32. V.V. Grushin, Application of the multiparameter theory of perturbations of Fredholm operators to Bloch functions. Mat. Zametki **86**(6), 819–828 (2009)
33. T. Kato, *Perturbation Theory for Linear Operators*, 2nd edn. (Springer, Berlin, New York, 1976). Grundlehren der Mathematischen Wissenschaften, Band 132
34. M. Kha, Green's function asymptotics of periodic elliptic operators on abelian coverings of compact manifolds. J. Funct. Anal. **274**(2), 341–387 (2018). https://doi.org/10.1016/j.jfa.2017.10.016
35. M. Kha, *A Short Note on Additive Functions on Riemannian Co-Compact Coverings* (2015). arXiv:1511.00185
36. M. Kha, V. Lin, *Polynomial-Like Elements in Vector Spaces with Group Actions Selim Grigorievich Krein Centennial*. Differential Equations, Mathematical Physics, and Applications: Selim Grigorievich Krein Centennial, pp. 193–217, 2019. https://doi.org/10.1090/conm/734/14772
37. M. Kha, P. Kuchment, A. Raich, Green's function asymptotics near the internal edges of spectra of periodic elliptic operators. Spectral gap interior. J. Spectr. Theory **7**(4), 1171–1233 (2017). https://doi.org/10.4171/JST/188
38. W. Kirsch, B. Simon, Comparison theorems for the gap of Schrödinger operators. J. Funct. Anal. **75**(2), 396–410 (1987)
39. F. Klopp, J. Ralston, Endpoints of the spectrum of periodic operators are generically simple. Methods Appl. Anal. **7**(3), 459–463 (2000)
40. T. Kobayashi, K. Ono, T. Sunada, Periodic Schrödinger operators on a manifold. Forum Math. **1**(1), 69–79 (1989)

41. P. Kuchment, *Floquet Theory for Partial Differential Equations*. Operator Theory: Advances and Applications, vol. 60 (Birkhäuser Verlag, Basel, 1993)
42. P. Kuchment, An overview of periodic elliptic operators. Bull. (New Series) Am. Math. Soc. **53**(3), 343–414 (2016)
43. P. Kuchment, Y. Pinchover, Integral representations and Liouville theorems for solutions of periodic elliptic equations. J. Funct. Anal. **181**(2), 402–446 (2001)
44. P. Kuchment, Y. Pinchover, Liouville theorems and spectral edge behavior on abelian coverings of compact manifolds. Trans. Am. Math. Soc. **359**(12), 5777–5815 (2007)
45. P. Kuchment, O. Post, On the spectra of carbon nano-structures. Comm. Math. Phys. **275**(3), 805–826 (2007)
46. P. Kuchment, A. Raich, Green's function asymptotics near the internal edges of spectra of periodic elliptic operators. Spectral edge case. Math. Nachr. **285**(14–15), 1880–1894 (2012)
47. H.B. Lawson Jr., M.-L. Michelsohn, *Spin Geometry*. Princeton Mathematical Series, vol. 38 (Princeton University Press, Princeton, NJ, 1989)
48. P. Li, *Geometric Analysis*. Cambridge Studies in Advanced Mathematics, vol. 134 (Cambridge University Press, Cambridge, 2012)
49. P. Li, J. Wang, Polynomial growth solutions of uniformly elliptic operators of non-divergence form. Proc. Am. Math. Soc. **129**(12), 3691–3699 (2001). https://doi.org/10.1090/S0002-9939-01-06167-6
50. P. Li, J. Wang, Counting dimensions of L-harmonic functions. Ann. Math. (2) **152**(2), 645–658 (2000). https://doi.org/10.2307/2661394
51. V.Ya. Lin, Y. Pinchover, Manifolds with group actions and elliptic operators. Mem. Am. Math. Soc. **112**(540), vi+78 (1994)
52. W. Lück, Survey on geometric group theory. Münster J. Math. **1**, 73–108 (2008)
53. V.G. Maz'ya, B.A. Plamenevskiĭ, Asymptotic behavior of the fundamental solutions of elliptic boundary value problems in domains with conical points. Boundary Value Problems Spectral Theory (Russian) **243**, 100–145 (1979)
54. S. Mizohata, *The Theory of Partial Differential Equations* (Cambridge University Press, New York, 1973). Translated from the Japanese by Katsumi Miyahara. MR0599580
55. P. Morandi, *Field and Galois Theory*. Graduate Texts in Mathematics, vol. 167 (Springer, New York, 1996)
56. J. Moser, M. Struwe, On a Liouville-type theorem for linear and nonlinear elliptic differential equations on a torus. Bol. Soc. Brasil. Mat. (N.S.) **23**(1–2), 1–20 (1992). https://doi.org/10.1007/BF02584809
57. N.S. Nadirashvili, Harmonic functions with a given set of singularities. Funktsional. Anal. i Prilozhen. **22**(1), 75–76 (1988). https://doi.org/10.1007/BF01077730
58. P.W. Nowak, G. Yu, *Large Scale Geometry*. EMS Textbooks in Mathematics (European Mathematical Society (EMS), Zürich, 2012)
59. R.G. Pinsky, Second order elliptic operators with periodic coefficients: criticality theory, perturbations, and positive harmonic functions. J. Funct. Anal. **129**(1), 80–107 (1995)
60. E. Randles, L. Saloff-Coste, Convolution powers of complex functions on $\mathbb{Z}^d$. Rev. Mat. Iberoam. **33**(3), 1045–1121 (2017). https://doi.org/10.4171/RMI/964
61. M. Reed, B. Simon, *Methods of Modern Mathematical Physics. IV. Analysis of Operators* (Academic Press, New York, London, 1978)
62. B. Riemann, Theorie der Abel'schen Functionen. J. Reine Angew. Math. **54**, 115–155 (1857). https://doi.org/10.1515/crll.1857.54.115
63. G. Roch, Ueber die Anzahl der willkurlichen Constanten in algebraischen Functionen. J. Reine Angew. Math. **64**, 372–376 (1865). https://doi.org/10.1515/crll.1865.64.372
64. L. Saloff-Coste, Analysis on Riemannian co-compact covers. Surv. Differ. Geometry **IX**, 351–384 (2004)
65. R.G. Shterenberg, An example of a periodic magnetic Schrödinger operator with a degenerate lower edge of the spectrum. Algebra i Analiz **16**(2), 177–185 (2004)
66. M.A. Shubin, Weak Bloch property and weight estimates for elliptic operators. Sminaire quations aux drives partielles (Polytechnique) (Unknown Month 1989), 1–20 (eng)

67. M.A. Shubin, Spectral theory of elliptic operators on noncompact manifolds. Astérisque **207**(5), 35–108 (1992). Méthodes semi-classiques, Vol. 1 (Nantes, 1991)
68. M.A. Shubin, L^2 Riemann-Roch theorem for elliptic operators. Geom. Funct. Anal. **5**(2), 482–527 (1995). https://doi.org/10.1007/BF01895677
69. È.È. Šnol′, On the behavior of the eigenfunctions of Schrödinger's equation. Mat. Sb. (N.S.) **42**(84), 273–286 (1957); erratum **46**(88), 259 (1957)
70. A. Sobolev, *Periodic Operators: The Method of Gauge Transform*. Lectures at the I (Newton Institute, 2015). https://www.newton.ac.uk/event/pepw01/timetable
71. T. Sunada, A periodic Schrödinger operator on an abelian cover. J. Fac. Sci. Univ. Tokyo Sect. IA Math. **37**(3), 575–583 (1990)
72. L.-C. Wang, Stability in Gromov-Shubin index theorem. Proc. Am. Math. Soc. **125**(5), 1399–1405 (1997)
73. C.H. Wilcox, Theory of Bloch waves. J. Anal. Math. **33**, 146–167 (1978)
74. S.-T. Yau, Nonlinear analysis in geometry. Enseign. Math. (2) **33**(1–2), 109–158 (1987)
75. S.-T. Yau, *Open Problems in Geometry*. Differential Geometry: Partial Differential Equations on Manifolds (Los Angeles, CA, 1990), 1993, pp. 1–28
76. M.G. Zaĭdenberg, S.G. Kreĭn, P.A. Kučcment, A.A. Pankov, Banach bundles and linear operators. Uspehi Mat. Nauk **30**(5(185)), 101–157 (1975)

Index

$A \underset{\sim}{\lesssim} B$, 35
A_N^p, 25
$A_{s,N}^p$, 35
$C_c^\infty(\mathcal{X})$, 13
$D^\pm$, 15
E_k, 3
$H^s(E_k)$, 3
$H_k^s(X)$, 3, 10
$L(\mu, P)$, 17
$L_k^2(X)$, 3
L°, 13
$L_p(\mu, A, N)$, 25
$V_N^p(A)$, 9
$V_N^p(A)$, 8
$V_N^p(X)$, 8
$V_N^\infty(A)$, 10
$V_{s,N}^p(A)$, 35
$V_{s,N}^p(X)$, 35
$\mathrm{Dom}_K P$, 17
$\mathrm{Dom}\, P$, 12
$\Gamma(\mathcal{X}, \mu, P)$, 18
$\Gamma_\mu(\mathcal{X}, P)$, 19
$\mathrm{Im}\, P$, 12
$\mathrm{ind}_\mu(P)$, 18
$\deg_P(\mu)$, 16
$\mathbb{T}^{*d}$, 2
$\mathcal{V}_\phi^p(A)$ as, 9
$\mathcal{D}'(\mathcal{X})$, 18
$\mathcal{E}^s$, 10
$\mathcal{L}(\mathbb{C}^{m_r})$, 24
$\mu = (D^+, L^+; D^-, L^-)$, 15
$\mu^{-1} := (D^-, L^-; D^+, L^+)$, 15
$\bar{\partial}$-operator, vii
Dom', 13

$\tilde{L}^\pm$, 15
$\tilde{P}^*$, 18
$\tilde{P}u$, 18
$\tilde{\Gamma}_\mu(\mathcal{X}, P)$, 19

AMS, ix
annihilator, 13
Assumption
   $\mathcal{A}$, 24
      strengthened, 28
Atiyah, vii
Avellaneda, viii, 1

band function, 55
Beltrami, vii
Bloch, 1, 5, 67
   function, 6, 67
   solution, 6
   variety, 5, 6
      real, 5
Boshernitzan, v
Brillouin, 2, 4
   zone, 2, 4
bundle
   cotangent, 2
   line, 3

Colding, vii
covering, vii, viii
   abelian, viii, 1, 23
   co-compact, viii
   nilpotent, vii

© The Author(s), under exclusive license to Springer Nature Switzerland AG 2021
M. Kha, P. Kuchment, *Liouville-Riemann-Roch Theorems on Abelian Coverings*,
Lecture Notes in Mathematics 2245, https://doi.org/10.1007/978-3-030-67428-1

of sub-exponential growth, 30

Dedekind, 74
  lemma, 74
Dirac, 60
  cone, 60
  point, 60
direct integral, 4
divisor, vii, 14
  degree, 15, 16
  inverse, 15
  periodic, 86
  point, 14
  rigged, 14, 15
    associated with, 15
  trivial, 29
domain, 2
  fundamental, 2, 3
duality, 2, 13, 19
  non-degenerate, 2

edge
  spectral
    generic, 28
edge of the spectrum, viii
eigenvalue
  principal, 63

Fermi, viii, 5
  energy, 5
  surface, viii, 5, 6
    empty, 30
    finite, 23
    real, 5
  variety, 5
finite difference, 6
  "twisted", 6
Floquet, 1, 4, 67
  function, 7, 67
    growth, 7
    of order N, 6
  multiplier, 69
  solution, 6
  transform, 4, 70
form
  bilinear, 2
  sesquilinear, 2
Fredholm, 14
  alternative, 29
  index, vii, 26, 33
    additivity, 19

operator, 14
property, 14
function
  additive, 67
  automorphic, 3
  Bloch, 6, 67
  Floquet, 6, 67, 85
  G-multiplicative, 63
  harmonic , vii

Gel'fand, vii
Gromov, vii, viii, 1, 12, 17
  theorem, 32
Gromov-Shubin theorem, 17
group, 1
  action, 1
    co-compact, 1
    free, 1
  character, 2
    unitary, 2
  deck, 2
    rank, 23
  discrete, 1
  finitely generated, 1
  rank, 1, 2
  virtually nilpotent, 32
  word metric on, 31

Hölder, 26
  conjugate, 26

index, 14

Kha, ix
Kuchment, viii, 1

Laplace, vii
Laplace-Beltrami operator, vii
Laplacian, viii
lattice, 2
  reciprocal, 2
Li, viii
Lin, F.-H., viii, 1
Liouville, vii, 6, 23
  property, viii, 23
  theorem, vii, 6, 7, 23, 30, 85
Liouville-Riemann-Roch
  equality, 27, 29, 86
  inequality, 27

manifold
  Riemannian
    of sub-exponential growth, 31
Maz'ya, vii, ix
measure
  Haar, 4
  Lebesque, 4
Minicozzi, vii
Moser, viii, 1

Nadirashvili, vii, viii, 1, 12, 17
Nadirashvili-Gromov-Shubin Riemann-Roch-
    type formula, 21
Nadirashvili-Gromov-Shubin theorem, 17
Nadirashvili-Gromov-Shubin version of the
    Riemann-Roch theorem, 12, 17
NSF, ix

operator
  adjoint, 2
  bounded below, 2
  difference, 85
  elliptic, 1, 2
  of divergent type, viii
  periodic, 1
  pseudodifferential, 73
  real, 25
  Schrödinger, 6, 58
  self-adjoint, 3
  smoothing, 73
  uniformly elliptic, 33
orbit, 1
  space, 1, 3

pairing
  bilinear, 13
parametrix, 73
perturbation
  theory, 3, 24
Pinchover, viii, ix, 1
Plamenevskii, vii
point
  diabolic, 60
polar, 13

quasimomentum, 2

Riemann, vii
Riemann-Roch
  equality, viii, 18, 20, 84

formula, 33
inequality, viii, 18
  extension of, 26
theorem, vii, 1, 18, 23, 85
Riesz, 10
  projector, 10, 24
Roch, vii

Schauder, 73
  estimate, 73
Schrödinger, 6, 58
Schwartz, 73
  kernel, 73
Shnol', 30
  strong property, 31
  theorem, 30
Shubin, v, vii–ix, 1, 12, 17
Simons Foundation, ix
Singer, vii
Sobolev
  space, 3, 58
space
  secondary, 15
spectral, 3
  band, 3, 4
  edge, viii
    non-degenerate, 57
  gap, 3, 4
spectral edge
  generic, 57
spectrum, 4
Struwe, viii, 1

theorem
  of Liouville type, vii
torus, 1
  $\mathbb{T}^{*d}$, 2

unique continuation, 9
  strong, 42
  weak, 43

von Neumann
  algebra, 86

Wang, J., viii

Yau, S.-T., vii
Yau's problem, vii

LECTURE NOTES IN MATHEMATICS   🐎 Springer

Editors in Chief: J.-M. Morel, B. Teissier;

Editorial Policy

1. Lecture Notes aim to report new developments in all areas of mathematics and their applications – quickly, informally and at a high level. Mathematical texts analysing new developments in modelling and numerical simulation are welcome.

   Manuscripts should be reasonably self-contained and rounded off. Thus they may, and often will, present not only results of the author but also related work by other people. They may be based on specialised lecture courses. Furthermore, the manuscripts should provide sufficient motivation, examples and applications. This clearly distinguishes Lecture Notes from journal articles or technical reports which normally are very concise. Articles intended for a journal but too long to be accepted by most journals, usually do not have this "lecture notes" character. For similar reasons it is unusual for doctoral theses to be accepted for the Lecture Notes series, though habilitation theses may be appropriate.

2. Besides monographs, multi-author manuscripts resulting from SUMMER SCHOOLS or similar INTENSIVE COURSES are welcome, provided their objective was held to present an active mathematical topic to an audience at the beginning or intermediate graduate level (a list of participants should be provided).

   The resulting manuscript should not be just a collection of course notes, but should require advance planning and coordination among the main lecturers. The subject matter should dictate the structure of the book. This structure should be motivated and explained in a scientific introduction, and the notation, references, index and formulation of results should be, if possible, unified by the editors. Each contribution should have an abstract and an introduction referring to the other contributions. In other words, more preparatory work must go into a multi-authored volume than simply assembling a disparate collection of papers, communicated at the event.

3. Manuscripts should be submitted either online at www.editorialmanager.com/lnm to Springer's mathematics editorial in Heidelberg, or electronically to one of the series editors. Authors should be aware that incomplete or insufficiently close-to-final manuscripts almost always result in longer refereeing times and nevertheless unclear referees' recommendations, making further refereeing of a final draft necessary. The strict minimum amount of material that will be considered should include a detailed outline describing the planned contents of each chapter, a bibliography and several sample chapters. Parallel submission of a manuscript to another publisher while under consideration for LNM is not acceptable and can lead to rejection.

4. In general, **monographs** will be sent out to at least 2 external referees for evaluation.

   A final decision to publish can be made only on the basis of the complete manuscript, however a refereeing process leading to a preliminary decision can be based on a pre-final or incomplete manuscript.

   Volume Editors of **multi-author works** are expected to arrange for the refereeing, to the usual scientific standards, of the individual contributions. If the resulting reports can be forwarded to the LNM Editorial Board, this is very helpful. If no reports are forwarded or if other questions remain unclear in respect of homogeneity etc, the series editors may wish to consult external referees for an overall evaluation of the volume.

5. Manuscripts should in general be submitted in English. Final manuscripts should contain at least 100 pages of mathematical text and should always include

   – a table of contents;
   – an informative introduction, with adequate motivation and perhaps some historical remarks: it should be accessible to a reader not intimately familiar with the topic treated;
   – a subject index: as a rule this is genuinely helpful for the reader.
   – For evaluation purposes, manuscripts should be submitted as pdf files.

6. Careful preparation of the manuscripts will help keep production time short besides ensuring satisfactory appearance of the finished book in print and online. After acceptance of the manuscript authors will be asked to prepare the final LaTeX source files (see LaTeX templates online: https://www.springer.com/gb/authors-editors/book-authors-editors/manuscriptpreparation/5636) plus the corresponding pdf- or zipped ps-file. The LaTeX source files are essential for producing the full-text online version of the book, see http://link.springer.com/bookseries/304 for the existing online volumes of LNM). The technical production of a Lecture Notes volume takes approximately 12 weeks. Additional instructions, if necessary, are available on request from lnm@springer.com.

7. Authors receive a total of 30 free copies of their volume and free access to their book on SpringerLink, but no royalties. They are entitled to a discount of 33.3 % on the price of Springer books purchased for their personal use, if ordering directly from Springer.

8. Commitment to publish is made by a *Publishing Agreement*; contributing authors of multiauthor books are requested to sign a *Consent to Publish form*. Springer-Verlag registers the copyright for each volume. Authors are free to reuse material contained in their LNM volumes in later publications: a brief written (or e-mail) request for formal permission is sufficient.

Addresses:
Professor Jean-Michel Morel, CMLA, École Normale Supérieure de Cachan, France
E-mail: moreljeanmichel@gmail.com

Professor Bernard Teissier, Equipe Géométrie et Dynamique,
Institut de Mathématiques de Jussieu – Paris Rive Gauche, Paris, France
E-mail: bernard.teissier@imj-prg.fr

Springer: Ute McCrory, Mathematics, Heidelberg, Germany,
E-mail: lnm@springer.com

Printed in the United States
By Bookmasters